T0342122

Power Systems

Power Systems

Fundamental Concepts and the Transition to Sustainability

Daniel S. Kirschen
University of Washington
Seattle, USA

This edition first published 2024

© 2024 John Wiley & Sons Ltd.

All rights reserved. No part of this publication may be reproduced, stored in a retrieval system, or transmitted, in any form or by any means, electronic, mechanical, photocopying, recording or otherwise, except as permitted by law. Advice on how to obtain permission to reuse material from this title is available at http://www.wiley.com/go/permissions.

The right of Daniel S. Kirschen to be identified as the author of the editorial material in this work has been asserted in accordance with law.

Registered Offices

John Wiley & Sons, Inc., 111 River Street, Hoboken, NJ 07030, USA

John Wiley & Sons Ltd, The Atrium, Southern Gate, Chichester, West Sussex, PO19 8SQ, UK

For details of our global editorial offices, customer services, and more information about Wiley products visit us at www.wiley.com.

Wiley also publishes its books in a variety of electronic formats and by print-on-demand. Some content that appears in standard print versions of this book may not be available in other formats.

Trademarks: Wiley and the Wiley logo are trademarks or registered trademarks of John Wiley & Sons, Inc. and/or its affiliates in the United States and other countries and may not be used without written permission. All other trademarks are the property of their respective owners. John Wiley & Sons, Inc. is not associated with any product or vendor mentioned in this book.

Limit of Liability/Disclaimer of Warranty

While the publisher and authors have used their best efforts in preparing this work, they make no representations or warranties with respect to the accuracy or completeness of the contents of this work and specifically disclaim all warranties, including without limitation any implied warranties of merchantability or fitness for a particular purpose. No warranty may be created or extended by sales representatives, written sales materials or promotional statements for this work. This work is sold with the understanding that the publisher is not engaged in rendering professional services. The advice and strategies contained herein may not be suitable for your situation. You should consult with a specialist where appropriate. The fact that an organization, website, or product is referred to in this work as a citation and/or potential source of further information does not mean that the publisher and authors endorse the information or services the organization, website, or product may provide or recommendations it may make. Further, readers should be aware that websites listed in this work may have changed or disappeared between when this work was written and when it is read. Neither the publisher nor authors shall be liable for any loss of profit or any other commercial damages, including but not limited to special, incidental, consequential, or other damages.

Library of Congress Cataloging-in-Publication Data applied for

Hardback ISBN: 9781394199501

Cover Design: Wiley

Cover Image: © zhengzaishuru/Shutterstock

Set in 9.5/12.5pt STIXTwoText by Straive, Chennai, India

Printed and bound by CPI Group (UK) Ltd, Croydon, CR0 4YY

C9781394199501_300524

Pour Penny, qui m'inspire à écrire.

Contents

Preface

Our students will be responsible for implementing fundamental changes in the nature and design of power systems. They will have to support the integration of rapidly increasing amounts of generation from renewable energy sources, manage a massive increase in the demand for electricity, and master new power electronic technologies while leveraging legacy assets. I wrote this textbook to help instructors prepare their students for these challenges.

To this end, besides covering some important classical topics such as the modeling of components, power flow, fault calculations, and stability, this book provides:
- A detailed discussion of the demand for electricity and how it affects the operation of power systems.
- An overview of the various forms of conventional and renewable energy conversion.
- A primer on modern power electronic power conversion.
- A careful analysis of the technical and economic issues involved in load generation balancing.
- An introduction to electricity markets.

To become effective actors in this transition, our students must not only become proficient in the use of analysis techniques but must also develop a profound understanding of how power systems actually work. The text therefore covers not only the "what" and the "how" but also the "why" of power system engineering. In particular, the issue of uncertainty is addressed on several occasions.

Some of the problems at the end of chapters aim to reinforce an intuitive understanding. Others require a computer solution or simulation. These can be performed using various open-source software packages. The data for these problems is available in various formats on the website https://github.com/Power-Systems-Textbook/TextbookSimulations.

While writing this book, I received help from many colleagues and students. Thanks to Baosen Zhang, June Lukuyu, Duncan Callaway, Ian Hiskens, and Sairaj Dhople for their suggestions. I am particularly grateful to Lane Smith, Aya Alayli, and Gord Stephen for setting up some of the simulation examples. Adam Greenhall, Yury Dvorkin, Yize Chen, and Ling Zhang lent a hand with other parts of the books. Barnaby Pitt and Ning Zhang created some of the figures of Chapter 2. Mike Mayhew, Ahmed Alhumoud, Nakseung Choi,

Evan Bowman, Rebecca Fairbanks, Sophia Votava, Jianing He, and Sophornnet Peou spotted typos and mistakes.

Most of all, thank you Penny for all your help and support over the years.

Seattle, WA   *Daniel S. Kirschen*
18 September 2023

Nomenclature

Diacritical marks

$\overline{X}$ denotes a complex number or phasor

$\overline{X}^*$ denotes the complex conjugate of a complex number

$|\overline{X}|$ or X denotes the magnitude of a phasor or complex number $\overline{X}$

$\overline{V}_{xy}$ phasor denoting the voltage from point x to point y

Subscripts

0	zero-sequence
1	positive-sequence
2	negative-sequence
a	air
B	base quantity
eff	effective
Im	imaginary
k	kinetic
LL	line-to-line voltage
LN	line-to-neutral voltage
m	mechanical
P	phase quantities
Re	real
ref	reference value
S	sequence quantities
w	wind or water

Superscripts

avg	average
crit	critical
e_i	elbow point i

inj	injected
max	maximum
min	minimum
nom	nominal
nl	no-load
pu	per unit
TH	Thevenin equivalent
TH, 0	zero-sequence Thevenin equivalent
TH, 1	positive-sequence Thevenin equivalent
TH, 2	negative-sequence Thevenin equivalent

Symbols

$\mathcal{L}$	Lagrangian function
λ	Lagrange multiplier
ρ_a	specific mass of air
ρ_w	specific mass of water
η	efficiency
ω	angular frequency
ω_m	mechanical angular frequency
Γ	profit
Φ	magnetic flux
A	area
C	cost
C_f	capacity factor
C_p	coefficient of performance
e	induced voltage or electro-motive force
E	energy
$\overline{E}$	internal electro-motive force phasor
f	frequency
$\overline{I}$	current phasor
$i(t)$	current in the time domain
m	mass
P	active power
p	number of magnetic poles
$p(t)$	instantaneous power
pf	power factor
Q	reactive power
$\overline{S}$	complex power
S	apparent power
T	torque
v	speed
$v(t)$	voltage in the time domain
$\overline{V}$	voltage phasor
Y	complex admittance
Z	complex impedance

Acronyms

BESS	battery energy storage system
BMS	battery management system
CCGT	combined cycle gas turbine
CHP	combined heat and power
DFIG	doubly fed induction generator
HVDC	high-voltage DC
KCL	Kirchhoff's current law
KVL	Kirchhoff's voltage law
LCOE	levelized cost of energy
LSE	load-serving entity
OCGT	open cycle gas turbine
OPF	optimal power flow
pu	per unit
PV	photovoltaic
PVC	poly vinyl chloride
PWM	pulse width modulation
rpm	revolution per minute
SoC	state of charge
TSO	transmission system operator
XLPE	cross-linked polyethylene

About the Companion Website

This book is accompanied by a companion website:

www.wiley.com/go/kirschen/powersystems

This website includes:

Solution Manual

1

Introduction

1.1 What is a Power System?

Electricity provides a clean, efficient, versatile, and economical way to deliver energy. Efficient generators have been designed to convert other forms of energy into electrical energy. Transmission lines carry large amounts of energy over long distances. Electric motors are efficient and make possible precise motion control. Electricity is also the only way to power electronic devices. However, there is one drawback to using electricity as an energy vector: storing significant amounts of energy in electrical form is not practical. Delivering significant amounts of electrical energy must therefore take place as a continuous process. The rate at which electrical energy flows (i.e., power) is the fundamental concept. Generating electric power from primary energy sources, transmitting it over long distances, and converting it into another form of power for a variety of end uses on an uninterrupted basis requires a set of devices working in a coordinated fashion. This is what we call a power system.

Power systems come in a wide range of sizes. The interconnections of Europe and North America connect thousands of generators to millions of consumers over vast meshed networks. At the other end of the scale, the smallest power system consists of a single generator converting energy from a primary source to power a single, local load. In this book, we will use examples from small systems to explain concepts, and we will also explain the techniques that engineers use to analyze and operate the largest power systems.

1.2 What are the Attributes of a Good Power System?

Three main objectives guide the planning, design, and operation of power systems:

Reliability: Because many aspects of modern life have become dependent on the availability of electric power, any interruption in its supply causes major damage or at least a significant nuisance. We expect the lights to come on when we flip the switch and sensitive industrial processes to complete without disruption, even as the system is exposed to random fluctuations or when some components fail.

Power Systems: Fundamental Concepts and the Transition to Sustainability, First Edition. Daniel S. Kirschen.
© 2024 John Wiley & Sons Ltd. Published 2024 by John Wiley & Sons Ltd.
Companion website: www.wiley.com/go/kirschen/powersystems

Sustainability: Like any other large-scale human activity, power systems have an impact on our environment, through the use of finite resources and the emission of greenhouse gases and other pollutants. To achieve sustainability, we must aim to generate 100% of our electrical energy from renewable sources.

Economy: Access to a cheap supply of electricity fosters economic growth and makes it possible for households to redirect their limited financial resources to other purposes.

Clearly, these objectives conflict. For example, improving reliability typically costs money; the cheapest sources of energy are not always sustainable; increasing the proportion of energy from renewable sources creates reliability issues. Finding solutions that optimally balance these objectives is the fundamental aim of power system engineering.

1.3 Structure of a Power System

1.3.1 Physical Structure

Except for the smallest applications, power systems connect multiple generators to multiple loads because this networked structure supports the desirable attributes described in the previous section. Supplying electric power from multiple generators increases reliability because when one of them fails, the other can compensate for this loss and, most of the time, ensure that no consumer is left without power. Connecting many energy sources to many loads also improves the overall economy of the system in two ways. First, as we will see in Chapter 2, aggregating loads reduces the amount of generation capacity that must be built to keep consumers supplied at all times. Second, this aggregated load varies substantially over the course of a day or a year. When this load is low, we can supply it from the most cost-effective generators in the system and use the more expensive ones only when they are needed. Finally, aggregating generation from renewable sources such as wind and solar over a wide area supports the sustainability goal. The production of these generators is driven by wind speed and cloud cover, factors that are uncontrollable but vary from location to location. Leveraging this diversity in output levels the overall renewable generation, which makes it easier to predict and match to the overall demand for electric power.

Most power systems operate with ac rather than dc because ac voltages can be easily or lowered using transformers. This ability to operate different parts of the system at different voltages reduces losses. To illustrate this fact, consider the simple circuit shown in Figure 1.1 and suppose that we want to supply a given resistive load P at a voltage V through a line of resistance R. The relation between the power supplied to the load, the current, and the load voltage is:

$$P_{\text{load}} = V.I \tag{1.1}$$

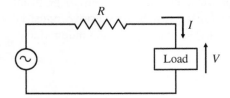

Figure 1.1 Simple circuit illustrating the effect of the supply voltage on the losses.

Table 1.1 Relative series losses for various standard operating voltages.

Voltage	Relative losses
110 V	1
220 V	0.25
13 kV	71.6×10^{-6}
132 kV	694.4×10^{-9}
345 kV	101.7×10^{-9}

The losses in the line are given by:

$$P_{\text{losses}} = I^2 R \tag{1.2}$$

Combining these two equations, we get:

$$P_{\text{losses}} = \frac{P_{\text{load}}^2}{V^2} R \tag{1.3}$$

The losses in the line are thus inversely proportional to the square of the operating voltage. Table 1.1 illustrates this effect for a few standard voltages used in power systems. Note that operating at very high voltages not only reduces losses but is also essential to maintaining the stability of the system. We will discuss this issue in Chapter 11.

Obviously, safety and practical considerations make it impossible to use very high voltages everywhere. To prevent accidents and accidental short circuits, conductors must be separated from each other and from ground by a distance or an amount of expensive insulating material that increases with the nominal operating voltage. While building tall towers in the countryside to support high-voltage transmission lines is feasible and economically justifiable, this is impossible in and around urban areas. Different parts of large power systems therefore operate at different nominal voltages, as illustrated in Figure 1.2. Design considerations typically limit the output voltage of large generators to less than 30 kV. This voltage is immediately stepped up to a voltage suitable for transmission over long distances. As the

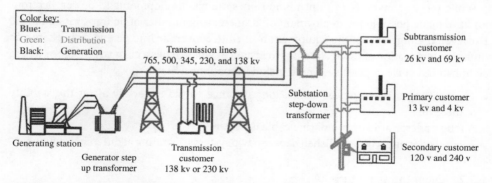

Figure 1.2 Typical operating voltages in large power systems. Source: United States Department of Energy, Blackout 2003, "Final Report on the August 14, 2003, Blackout in the United States and Canada: Causes and Recommendations".

amount of power and the distance over which it must be transmitted increase, so does the most appropriate standardized nominal voltage. This voltage is gradually stepped down to meet the needs of various classes of consumers. According to the U.S. Energy Information Agency, on average only about 5% of the electricity generated in the United States is dissipated as losses in the transmission and distribution networks. Most of these losses occur in the low-voltage distribution networks.

Very large consumers (for example steel processing plants or refineries) receive power directly at transmission voltage level and operate their own industrial power system. Smaller industrial, commercial, and residential consumers are connected at the lower voltage appropriate to their needs. The higher-voltage components constitute the transmission network, while the lower-voltage parts form the distribution network. An intermediate portion is sometimes described as the subtransmission network. Local conventions rather than universally agreed voltage thresholds define the demarcation between distribution, subtransmission, and transmission.

Transmission networks are generally meshed to further the triple objective of reliability, economy, and sustainability. Meshes indeed provide redundant paths for the flow of power from generators to loads. These redundant paths reduce the likelihood that this flow would be interrupted by the disconnection of a line. They also increase the transmission capacity, which is the maximum amount of power that cheap and renewable sources can securely provide over the network. With the exception of some densely populated urban areas, distribution networks are typically radial because protecting against the consequences of faults is easier and cheaper in a radial network than in a meshed network. However, radial networks do not provide the same level of reliability as meshed networks because the disconnection of a single component can disconnect consumers from all sources of power. This is considered acceptable in distribution networks because each occurrence of such a problem affects only a relatively small number of consumers.

Figure 1.2 reflects the operation of traditional power systems where power produced by a relatively small number of large, centrally controlled generators flows through the transmission network and down the radial distribution network to the consumers. In recent years, many small generators (typically photovoltaic) have been connected to distribution networks.

While the vast majority of transmission lines are ac, developments in power electronics have made possible the deployment of an increasing number of dc transmission lines. However, the high cost of the equipment needed to convert ac into dc and dc back into ac limits the application of this technology to instances where it has a clear advantage or where ac transmission is not possible. These include situations where:

- Transmitting a lot of power over a long distance would adversely affect the stability of the grid.
- A long underground or submarine cable is required.
- A connection must be established between two systems operating at different frequencies.

1.3.2 Cyber Infrastructure

While power plants and transmission lines are the most visible part of power systems, these systems would not operate reliably or economically without the support of an extensive set of communications, control, and computing devices. This infrastructure includes of the following types of devices:

- *Control systems* that regulate the operation of active devices, such as power plants, generators, power electronics converters and transformers, to ensure that critical variables remain close to their setpoint value.
- *Meters* that record voltages, currents, powers, and other variables for monitoring or billing.
- *Relays* that detect various types of faults on the physical components and trigger actions to avoid further damage.
- *Control centers* where human operators oversee the operation of a power plant or the overall system.
- *Communication networks* that transmit the data collected by the meters to the control centers and allow the operators to remotely control devices across the system.
- *Computers* to support the monitoring of the system, optimize its operation, and model what might happen under various scenarios.

1.3.3 Organizational Structure

Historically, most power systems were built and operated by a single entity: the electric utility. This company or government agency owned all the assets and controlled every aspect of the system's operation and development. In many parts of the world, this vertically integrated structure has been unbundled to make possible the introduction of competitive electricity markets. In this new structure, generating companies compete to sell electrical energy on a wholesale market and retailers compete to resell this energy to individual consumers. Usage of the transmission and distribution and distribution networks is shared by all participants under the watchful eye of a system operator who is responsible for maintaining the reliability of the system. Chapter 12 discusses in more detail how electricity markets function and the coordination between these various entities. In the meantime, we need not be concerned about who owns and operates each part of the power system. For simplicity, we will use the generic term "utility" to refer to the entity in charge of the part of the system under consideration.

1.4 Historical Evolution

As electricity started being used as a source of light and power in the late 19th century, power systems consisted of a few wealthy customers supplied by a single power plant in their neighborhood. Thomas Edison championed dc, while George Westinghouse and Nikola Tesla promoted ac. As demand grew and power systems needed to reach more widely spread customers and rely on more remote sources of primary energy, the advantages of ac became clear, and dc quickly fell out of favor. Electric utilities companies focused their efforts on urban areas because it is more profitable to supply customers when they are located close together rather than when they are widely spaced. Rural electrification required the emergence of electricity cooperatives and the provision of government subsidies.

Alaska
interconnection

Quebec
interconnection

Western
interconnection

Eastern
interconnection

Texas
interconnection

Figure 1.3 Interconnected areas in North America. Source: Congressional Budget Office/https://
www.cbo.gov/publication/56254/Public Domain.

Utilities quickly realized the economic and reliability benefits of interconnecting their
systems. One company might have excess power capacity that it could sell to the other at a
price that was advantageous for both. Each company could also provide backup power for
the other in case of need. Over time, these interconnections involved more companies and
became larger and larger. As Figures 1.3 and 1.4 show, the largest interconnections span
continents, link thousands of generators, and supply electrical energy to tens of millions of
consumers.

As the demand for electricity grew, economies of scale encouraged the construction of
larger and larger power plants, often located close to remote and non-transportable sources
of primary energy, such as hydroelectricity, wind, and solar. In parallel, this required the
erection of transmission lines at higher and higher nominal voltages. In recent years, a
slight reversal has occurred with the development of distributed energy resources, primar-
ily solar photovoltaic (PV) generation and battery energy storage, that can be deployed
cost-effectively at smaller scale. These resources are connected to the distribution network
rather than the transmission network. Occasionally, when their production exceeds the
local load, the flow of power in the local distribution network reverses direction, i.e., power
flows from the distribution network back into the transmission network. While the deployment of distributed PV generation on homes and businesses allows these consumers to
reduce their electricity bills, it is unlikely to completely displace centralized generation, as

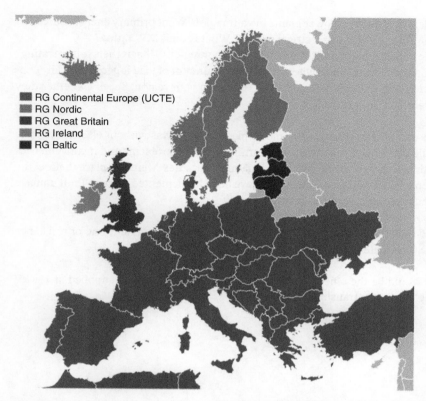

RG Continental Europe (UCTE)
RG Nordic
RG Great Britain
RG Ireland
RG Baltic

Figure 1.4 Interconnected areas in Europe. Source: Kimdime/Wikimedia Commons/Public Domain.

the economies of scale still favor larger plants and the density of solar energy is insufficient to meet the needs of urban areas.

A microgrid is a small portion of the grid that includes some form of generation and sometimes an energy storage device, and which has the ability to separate itself from the main grid and continue operating independently for some time. A microgrid is thus able to serve some loads if the main grid were to fail due to either a natural disaster, such as an earthquake or major storm, or a blackout triggered by an electrical fault. Microgrids have been attracting a growing amount of attention because they provide a way to enhance the resilience of critical facilities and communities.

Problems

P1.1 Besides electricity, what other energy delivery systems are in common use? Discuss their advantages and disadvantages compared to power systems.

P1.2 Obtain a map showing the configuration of your regional or national power system.
(a) Describe the geographical area covered by this power system.

(b) Identify the major power plants in your region. What primary energy sources do they use to generate electrical energy? What is their MW rating?

(c) Identify the major transmission lines in your region. What is their voltage rating?

(d) Load centers are areas where a significant amount of load is geographically concentrated. Identify the major load centers in your region. Specify whether they are urban or industrial.

P1.3 When was the last time your home or place of work was without electricity? How long did it last? Did you have prior warning? List the problems and inconveniences that this outage caused. What mitigation measures were in place before the incident? What mitigation measures have been implemented since to limit damage and inconvenience during a future outage?

P1.4 A load of 100 MW is supplied at 230 kV through a transmission line of resistance 5.29 Ω. Calculate the efficiency of this transmission.

P1.5 What would be the efficiency if the load of problem P1.4 were supplied at 138 kV through the same transmission line?

2

Electrical Loads and the Demand for Electricity

2.1 Overview

Loads is the term used to describe the components of the system that convert electrical energy to other forms of energy and in the process deliver value to the consumers. In a power system, these components are controlled by human beings and thus reflect their demand for electricity.

This chapter discusses the physical modeling of the loads, and how loads are affected by the behavior of consumers. It also considers some financial aspects of the relation between these consumers and their utility company.

2.2 Residential Loads

Figure 2.1 shows a simplified diagram of the wiring of a house. Since the utility is required to maintain the supply voltage within a relatively narrow band, we will represent the connection of this house to the power system by an ideal voltage source. Closing a switch connects a load (a light, a refrigerator, a computer, etc.) to this voltage source. All the loads in this house are connected in parallel. In fact, all of the loads in a power system are connected in parallel with all the generators in the system.

It would be tempting to model these loads using their impedance because this would allow us to use directly all the techniques of classical circuit analysis. In practice, however, loads are specified in terms of their power consumption. Power is indeed what the meter outside the house records because consumers get charged as a function of the amount of energy that they consume. Table 2.1 shows typical values of the power ratings of some appliances commonly found in houses. The rating reflects the power consumption of the appliance when it is fully on and operating at nominal voltage.

Figure 2.2 shows the power consumption profile of four houses, recorded every minute over the course of the same day. While these profiles look different, they exhibit similar patterns:

- A very small constant load
- Fairly regular, periodic increases probably due to thermostatically controlled appliances such as refrigerators

Power Systems: Fundamental Concepts and the Transition to Sustainability, First Edition. Daniel S. Kirschen.
© 2024 John Wiley & Sons Ltd. Published 2024 by John Wiley & Sons Ltd.
Companion website: www.wiley.com/go/kirschen/powersystems

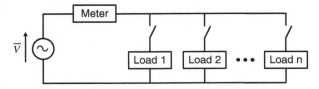

Figure 2.1   Simplified wiring diagram of a house.

Table 2.1   Typical rating range of some appliances.

Appliance	Typical rating
LED lightbulb	9–15 W
Incandescent lightbulb	60–100 W
Laptop computer	65–100 W
Desktop computer	250–500 W
Refrigerator	300–1100 W
Single-room air conditioner	600–1500 W
Space heater	750–1500 W
Water heater with tank	2000–6000 W
Whole-house air conditioner	2500–4000 W
Electric shower	8000–10,000 W
Tankless water heater	13,000–36,000 W

- Morning, midday, and evening periods of increased demand when occupants are more likely to be home and actively using appliances
- Short spikes in demand when large appliances are turned on.

Figure 2.3 illustrates what happens if we aggregate the load of 16 houses (including the four shown in Figure 2.2):

- The constant load is much more substantial.
- The morning, midday, and evening periods of higher demand ramp up and down more smoothly, reflecting that different people are on different schedules.
- As Table 2.2 shows in more detail, the ratio of the peak to the average load is considerably smaller for the aggregated demand than it is for any of the four houses shown in Figure 2.2.

If we assume that the peak loads on that day are representative of the annual peak loads of these houses and that each of these 16 houses had to install a generator sufficiently large to meet its own peak load, a total of about 95 kW of generation capacity would have to be deployed, instead of about 25 kW to meet the aggregated peak load.

The capacity factor of a generator is defined as the ratio of the actual amount of energy that it produces over a given period to the amount of energy that it could produce if it produced at maximum capacity at all times during that same period:

$$C_f = \frac{\int_0^T P(t)dt}{T \times P^{max}} \tag{2.1}$$

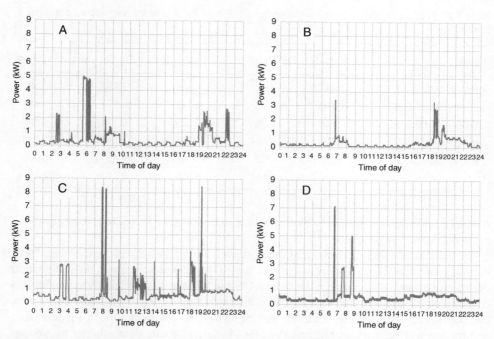

Figure 2.2 Power consumption profiles of four houses, recorded every minute over the course of the same day.

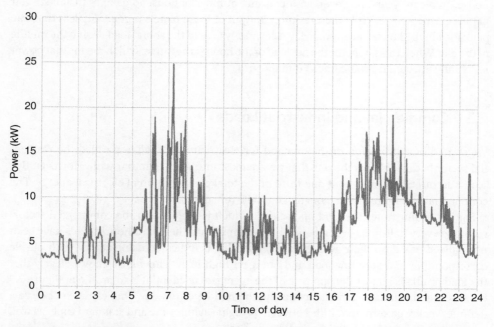

Figure 2.3 Aggregated power consumption profile of 16 houses, recorded every minute over the course of the same day.

Table 2.2 Features of the four residential profiles of Figure 2.2 and of the aggregated profile of Figure 2.3.

House	A	B	C	D	16 houses
Peak load (kW)	5.013	3.413	8.464	7.157	24.887
Minimum load (kW)	0.045	0.083	0.141	0.179	2.454
Average load (kW)	0.507	0.3027	0.7398	0.5747	6.985
Energy (kWh)	12.171	7.264	17.754	13.794	167.633
Peak/average ratio	9.88	11.27	11.44	12.45	3.56
Capacity factor	10.1%	8.9%	8.7%	8.0%	28.1%

where $P(t)$ is the power produced by a generator at time t and P^{max} is the capacity or maximum output of this generator.

The last line of Table 2.2 shows that the capacity factor of these fictitious individual residential generators would be at or below 10%, which would represent a very poor utilization of expensive assets. Aggregating the demand of the 16 houses significantly improves this capacity factor because it takes advantage of the *diversity* in the demand of these consumers, i.e., the fact that they all use electricity according to their own schedule. In the following section, we will see that aggregating the demand for electricity over a larger number of consumers of different types amplifies this effect.

None of the consumers in this small sample had actually installed a generator in their house. In recent years, however, an increasing number of households have installed solar panels on their roof. While the maximum output of these panels is generally less than the peak load in the house, on a sunny day it can easily exceed the actual load during the middle of the day. When this happens, the net load of the house is negative, which means that power is actually injected from the house into the grid.

2.3 Commercial and Industrial Loads

We started our discussion of the demand for electricity with what is called the residential sector because that is the one that we are most intuitively familiar with. However, this sector accounts for only about one-third of the total consumption of electrical energy. The commercial and industrial sectors make up the remainder.

Figure 2.4 illustrates the load of a school (which is included in the commercial sector) recorded every half-hour over a period of seven months. These half-hourly loads have been normalized based on their average and their value is indicated using colors, with purple corresponding to a very low load and orange a high load. This figure shows clearly that the load profile is almost the opposite of the aggregate residential load profile of Figure 2.3 because the maximum load occurs during the middle of the day. One can also observe a definite weekly pattern, the shifts between daylight savings time and standard time, as well as a change in pattern around the holiday period.

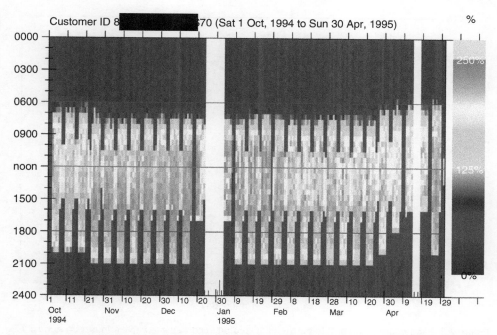

Figure 2.4 Power consumption of a school, recorded every half-hour over the course of seven months. The color hue indicates the value relative to the average load. The vertical white bands correspond to periods of missing data.

The load profiles for industrial consumers are very diverse. Process industries have very flat profiles because they run continuously. The pattern for other industries depends on how many shifts they employ and whether their production involves large intermittent loads.

2.4   Load Aggregation over a Large Region

Aggregating loads over a large and diverse group of consumers yields a profile that is considerably smoother than the individual consumer profiles shown in Sections 2.2 and 2.3. For example, Figure 2.5 shows the aggregated profile of the loads supplied on August 12, 2020, by the Midcontinent Independent System Operator (MISO), which serves 42 million consumers in the central part of the United States and Canada. On this day, the maximum load was 104,347 MW and the average load was 85,898 MW, giving a ratio of maximum to average of only 1.21. This aggregated load profile is not only smoother and flatter than the load profile of an aggregation of a smaller number of consumers, it is also much easier to predict accurately because individual decisions about when to connect a load have a negligible effect on the total.

As we will discuss in Chapter 6, serving flatter, smoother, and more predictable load profiles is significantly easier, cheaper, and less likely to result in outages.

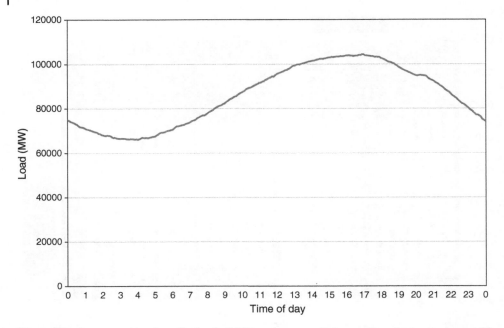

Figure 2.5 Aggregated load profile for the MISO region, recorded every five minutes on August 12, 2020. Source: Adapted from www.misoenergy.org.

2.5   What Factors Shape the Aggregated Load Profile in the Short Run?

Astronomical and meteorological factors have a synchronizing effect on the behavior of consumers of electricity. Astronomical factors shape the aggregated load profile and alter it from day to day:

Time of day: The aggregated load increases as more people get out of bed and go to work. It often dips a bit during the afternoon before rising again in the evening as people cook dinner, do laundry, and watch television. It then gradually decreases and reaches a minimum in the early hours of the morning before the cycle starts again.

Day of the week: Profiles on weekend days are significantly different from profiles during weekdays because industrial and commercial loads decrease while residential loads increase. There are also less pronounced differences between the weekdays.

Time of the year: The need for electric lighting is directly influenced by the time at which the sun rises and sets. While this effect is gradual, switching between daylight savings time and standard time introduces a sudden shift.

Holidays have similar effects as weekends, albeit more pronounced.

Meteorological factors affect the load profile in the following ways:

Temperature has a very significant effect on the electrical load when it deviates from the human comfort zone. In warmer climates, the electrical load increases during the summer months as people use more air conditioning. In colder climates, the inverse happens when people rely on electric heating. Figure 2.6 illustrates these effects for two large power systems, one with cold winters (PJM), the other with hot summers (ERCOT).

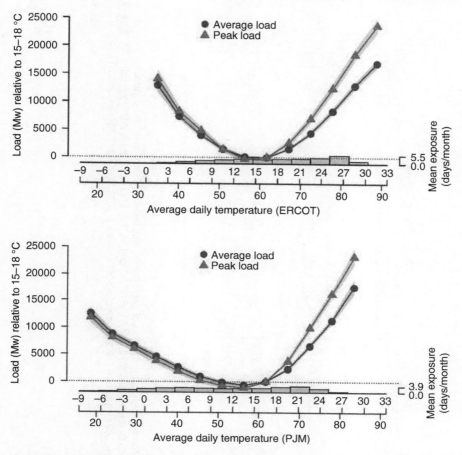

Figure 2.6 Changes in the average and peak load as a function of the average daily temperature for the ERCOT and PJM systems. Source: Kimdime/Wikimedia Commons/Public Domain.

Humidity amplifies the effects of high and low temperatures.
Wind increases heat losses from buildings and thus increases the need for electric heating.
Cloud cover increases the lighting load.

Figure 2.7 illustrates these astronomical and meteorological effects by showing the load profiles of two systems on the days with the highest and lowest loads. In Great Britain, the peak load typically occurs on a very cold weekday during the darkest months of winter, while the smallest load occurs on a Sunday with a pleasant temperature. On the other hand, in Texas, due to the heavy use of air conditioning, the peak load typically occurs on a hot, humid summer day, while the smallest load occurs in spring or autumn when the temperature is neither too hot nor too cold.

Because system operators have to procure and schedule the resources needed to reliably meet the load, they must forecast this load profile at least one day ahead. To this end, they use archived time series of load data and meteorological variables to develop sophisticated forecasting models that are able to predict the load with an accuracy of a few percent.

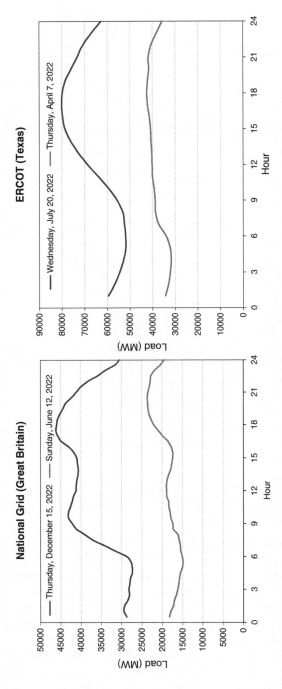

Figure 2.7 Load profiles on the days with the highest and lowest loads in the systems of Great Britain and Texas. The maximum and minimum temperatures in London were 36 °F (2 °C) and 21 °F (−6 °C) on December 15, 2022, and 81 °F (27 °C) and 55 °F (13 °C) on June 12, 2022. In Austin, TX these temperatures were 103 °F (40 °C) and 74 °F (23 °C) on July 20, 2022 and 76 °F (24 °C) and 42 °F(6 °C) on April 7, 2022. (Data: NGCESO and ERCOT).

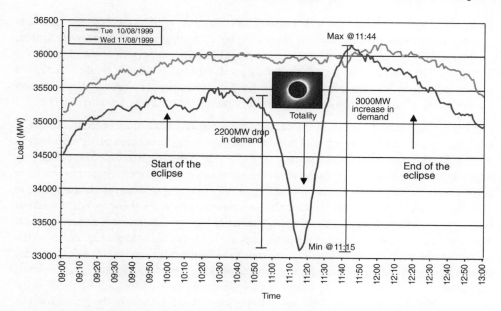

Figure 2.8 Effect of the total solar eclipse of August 11, 1999, on the electrical load of Great Britain. Source: National Grid UK/CC BY-ND 4.0.

Sporting and other special events of particular significance that are broadcast on television, tend to synchronize the activities of a large part of the population. For example, at the end of a football game, many people will turn off their television or switch on the kettle to make a cup of tea. This loss of diversity can create significant swings in the electrical load of the system. Figure 2.8 illustrates how this synchronization affected the load in Great Britain during the total solar eclipse of August 11, 1999. For comparison, this figure shows the load profile on the previous day. Because events of this type are rare, their effect on the load profile is very difficult to forecast.

2.6 What Affects the Aggregated Electrical Load in the Long Run?

Historically, the consumption of electrical energy has been closely linked to economic activity. In particular, during the second half of the twentieth century, as industry and commerce flourished, so did the demand for electricity. Conversely, developments in electrical technologies spurred economic growth. At the same time, as the income of households grew, they purchased more electrical appliances and consumed more electrical energy. Until about the year 2000, graphs tracking the demand for electrical energy followed closely the evolution of the gross domestic product (GDP). Since then, the rate of growth in the demand for electricity has been considerably smaller than the rate of economic growth. Three factors are probably responsible for most of this decrease in the electric energy intensity of the economy:

Technological advances: power electronics, variable speed motor drives, and many other new sensing and control technologies make it possible to produce more goods while consuming less electricity.

Energy efficiency standards: the stricter building codes and appliance efficiency standards that governments have enacted are starting to have an effect as buildings are renovated, older appliances are replaced, and incandescent lighting is superseded by LEDs.

Structural shift in the economy: manufacturing goods represents a much smaller fraction of the economy in Western countries than it used to. It has been replaced by the provision of services, which consumes considerably less energy for the same amount of GDP.

Previous long-term forecasts of the demand for electricity have often been inaccurate, if not grossly incorrect. Predicting how this demand will change over the coming years is even more difficult because of some new factors. First, a deep decarbonization of the economy will compel a reduction in the use of fossil fuels across all sectors, including transportation and heating. Achieving this reduction will require converting this energy consumption to electricity, which will not only increase the total amount of electrical energy consumed but could also substantially affect the shape of the daily load profile. Second, as summers get hotter because of climate change, the peak summer loads are likely to get higher due to the increased use of air conditioning. Regions that have historically seen their peak electricity demand occur in winter might become summer peaking. Third, as the cost of solar panels decreases, more consumers will install such systems on their premises, i.e., "behind the meter." While the sun shines, these installations provide at least part of the power used by these consumers and thus reduce the net load that the rest of the system must provide. On an aggregated basis, these solar panels will significantly modify the shape of the system load curve. For example, taking as a reference the load profile for a spring day in 2014 before any significant amount of behind the meter solar generation was installed, Figure 2.9 shows the

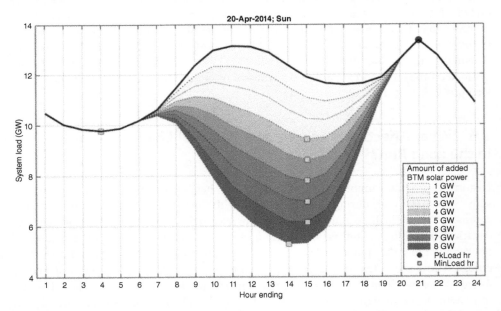

Figure 2.9 Effect of increasing amounts of behind the meter solar generation capacity on the shape of the system load for a typical spring day in the New England region. Source: ISO New England/https://www.iso-ne.com/static-assets/documents/2017/05/clg_meeting_black_panelist_presentation_june_1_2017_final.pdf.

effect that ISO-New England expects increasing amounts of installed solar capacity would have on the shape of its load profile.

2.7 Metering and Billing

For most residential and small commercial consumers, the meter shown in Figure 2.1 integrates over time the active power flowing into the loads:

$$E = \int_0^T \left\{ \sum_{i=1}^N P_i(t) \right\} dt \tag{2.2}$$

where T is the billing interval, which is usually one month. The company selling electricity to this consumer then sends them a bill that is typically calculated as follows:

$$A = E \times c + FC \tag{2.3}$$

where A is the amount to be paid (\$), c is a tariff rate in \$/kWh, and FC (\$) is a small, fixed charge.

   This billing formula is simple and encourages consumers to reduce their energy consumption, which is desirable from a sustainability perspective. On the other hand, it does not reflect the cost to the utility of supplying electricity to the consumers. A small fraction of the utility's cost is proportional to the amount of fuel needed to produce the electrical energy delivered. A much larger fraction reflects the cost of building the infrastructure required to generate and deliver this energy, i.e., power plants, transmission lines, and distribution networks. This second component is not proportional to the amount of energy but to the aggregated peak load of the utility. Furthermore, as we will see in Chapter 6, the cost of producing electrical energy varies over time. Factoring this time dependence in the bill would be confusing for most residential consumers and would require the installation of more sophisticated and costly meters.

   The billing formula for larger commercial and industrial consumers is generally more complex and factors in their peak load.

Example 2.1 *An Electricity Bill*   Figure 2.10 shows the load profile of an industrial consumer. Calculate the electricity bill for this day assuming that it is charged 0.12 \$/kWh for energy, that it pays 0.75 \$/kW for its peak load, and that the fixed charge is \$15.

Solution
Total energy consumed:

$$E = 500 \times 15 + 1000 \times 3 + 2000 \times 2 + 1500 \times 4 = 20,500 \text{ kWh}$$

Peak load:

$$P^{max} = 2000 \text{ kW}$$

Daily electricity bill:

$$A = 20,500 \times 0.12 + 2000 \times 0.75 + 15 = \$3975.00$$

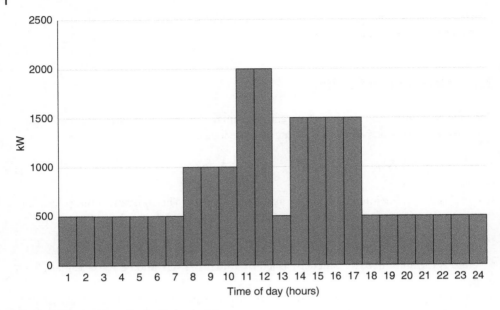

Figure 2.10 Load profile for Example 2.1.

2.8 Flexibility

Since the days of Thomas Edison, electricity has been marketed on the basis of its convenience: by flipping a switch, consumers can get as much power as they want, when they want it, and as long as they want it without having to store it. If consumers are charged for this electricity according to the constant per kWh rate formula discussed in Section 2.7, they have no incentive to modulate their demand. The aggregated load profiles therefore follow the cycles of human activity. Having promised convenience, companies selling electricity must meet this load. However, tracking the ups and downs of the load profile is costly because the generating plants that are built to supply the peak load sit idle during the troughs. The increasing proportion of generation from renewable sources such as wind and solar exacerbates this issue because their production is often not correlated with the demand for electricity. Managing large and rapid changes in the load can be challenging, which could affect the reliability of the supply.

These concerns about cost and reliability have led many utilities to explore how they might convince consumers to be more flexible in their demand for electricity. Being flexible means being willing to reduce one's demand or, more commonly, shifting part of the demand from a peak load period to an off-peak period. Flexibility depends very much on the type of load. It is pointless to ask consumers to turn off their television during a major sporting event. At the other extreme, a shift in electricity demand will go unnoticed when the electrical energy is used for processing something that can be stored, such as heating hot water, washing clothes, cooling a building, or manufacturing intermediate products. Because they have become accustomed to getting electricity at will, consumers need a financial incentive to be more flexible in their demand for electricity.

For example, some industrial consumers buy electricity on the basis of interruptible contracts. In exchange for a lower tariff, such contracts give the supplier the right to force the consumer to temporarily reduce its load a few times a year at short notice. As we will discuss in Chapter 12, the development of electricity markets has clarified the relation between the load profile and the cost of supplying electricity, and thus highlighted the benefits of flexibility. Nevertheless, the fraction of the overall demand that is actually flexible remains small because the value that consumers place on a continuous supply of electricity is typically much higher than the price that they pay. For example, the cost to a manufacturer of rescheduling its production at short notice may be much larger than the financial incentive that its electricity supplier might be willing to provide.

2.9 Outages

Occasionally, some consumers are disconnected from the power supply without notice or prior agreement. This happens either because one or more components of the network has failed, or because it is necessary to reduce the load to stabilize the system. Depending on the nature of the problem, these outages can last from a few seconds to a few days and are extremely disruptive in societies that have become heavily dependent on a continuous supply of electric power. Consumers affected by such outages risk injuries or death, incur financial losses, and suffer possibly severe inconvenience.

Quantifying in monetary terms how much damage these outages cause is useful because it helps determine how much money ought to be spent trying to reduce their probability of occurrence. However, this is not an easy task because the human and economic impacts of an outage depend on a number of factors:

- *The vulnerability of the individuals*: While college students might not be bothered too much, frail older adults, particularly those who depend on electrically powered medical devices, can be severely affected.
- *The nature of the business*: Would an outage cause the shutdown of an industrial process that requires hours to restart? Would it damage manufactured products or soil foodstuff?
- *The timing of the outage*: Does it occur in the middle of the night, during working hours, at rush hour, or in the middle of the Super Bowl?
- The duration of the outage.
- The geographical extent of the outage.

The most common way to carry out this quantification is to survey a representative number of individuals and companies, and ask them to estimate the costs of various outage scenarios or to state how much they would be willing to pay to avoid such outages. The results of these surveys are often summarized using a quantity called the Value of Lost Load (VoLL), which is defined as how much an average consumer would be willing to pay to avoid being disconnected without warning, and not being able to consume 1 kWh of electrical energy. Table 2.3 summarizes the results of such studies for various countries and regions. The disparities between the values for similar countries suggest that VoLL is a very rough metric of the potential impact of outages. However, even if we accept that these values are inaccurate, they suggest that on average the value that consumers put

Table 2.3 Value of Lost Load (VoLL) from various countries and regions.

Country/Region	VoLL ($/MWh)
United Kingdom	$22,000
Countries of the European Union	$12,290–$29,050
United States	$7500
MISO	$3500
New Zealand	$41,269
Victoria – Australia	$44,438
Australia	$45,708
Ireland	$9538
Northeast USA	$9283–$13,925

Sources: Adapted from London Economics, European Commission, Brattle Group, MISO. Where necessary, the values have been converted to US dollars.

on not being disconnected is about 2 orders of magnitude larger than what they pay for electrical energy.

It is important to note that VoLL pertains to interruptions without notice and that it is an average over all the loads of all the consumers. Some consumers value some of their loads at much less than these average values, which explains why they are willing to occasionally disconnect them temporarily as discussed in Section 2.8.

2.10 Complex Power, Reactive Power, and Power Factor

When an ac voltage is applied to some loads, such as lighting and heating, the current drawn by these loads is in phase with the voltage. If we represent these quantities as phasors and choose the voltage as the reference for the angles, we have:

$$\overline{V} = V \angle 0°$$
$$\overline{I} = I \angle 0°$$

(2.4)

where the magnitudes are expressed using RMS values.

For other loads, such as those that involve motors, the current typically lags the voltage and we have:

$$\overline{V} = V \angle 0°$$
$$\overline{I} = I \angle - \theta°$$

(2.5)

The *complex power* drawn by a load is defined as:

$$\overline{S} = \overline{V}\overline{I}^{*} = V\angle 0° I\angle \theta° = VI(\cos\theta + j\sin\theta) = VI\cos\theta + jVI\sin\theta$$

(2.6)

Or

$$\overline{S} = P + jQ$$

(2.7)

Figure 2.11 Power triangle for lagging and leading power factors.

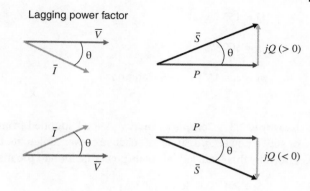

where:

$P = VI \cos \theta$ is the *active power* consumed by the load, i.e., the rate at which electrical energy is converted to another form of energy and put to practical use. Its units are W, kW, or MW.

$Q = VI \sin \theta$ is the *reactive power* drawn by the load. Reactive power represents power that oscillates in the system at twice the frequency of the voltage without being dissipated or put to practical use. While dimensionally identical to the active power, it is measured in var (Volt Ampere reactive), kvar, or Mvar to emphasize the distinction.

The apparent power is the magnitude of the complex power:

$$S = |\bar{S}| = \sqrt{P^2 + Q^2} = VI \tag{2.8}$$

It is measured in VA (Volt Ampere), kVA, or MVA.
The *power factor* is defined as follows:

$$pf = \frac{P}{\sqrt{P^2 + Q^2}} = \frac{P}{S} = \cos \theta \tag{2.9}$$

When the current drawn by a load lags the voltage, the reactive power is positive, and the power factor is said to be lagging. On the other hand, when the current leads the voltage, the reactive power is negative, and the power factor is said to be leading. Figure 2.11 illustrates these two cases. Power engineers like to say that inductive loads "consume" reactive power, while capacitive loads "produce" reactive power.

Example 2.2 *Complex Power* A load is supplied by a 120 V source and draws a current of 5 A, which lags the voltage by 15°. Calculate the complex, active, reactive and apparent power supplied to this load, as well as its power factor.

$$\bar{V} = 120\angle0°$$

$$\bar{I} = 5\angle-15°$$

$$\bar{S} = \bar{V}\bar{I}^* = 120\angle0° \times 5\angle15° = 579.55 + j155.29 \text{ VA}$$

$$P = 579.55 \text{ W}$$

$$Q = 155.29 \text{ var}$$

$$S = 600 \text{ VA}$$

$$pf = \cos(15°) = 0.966 \text{ lagging}$$

Example 2.3 *Complex Power* A load supplied from a 240 V source consumes 100 W of active power and 50 var of reactive power. Calculate the complex and apparent powers drawn by this load, as well as its power factor and the magnitude and phase of the current.

$$\overline{S} = P + jQ = 100 + j50 \text{ VA}$$

$$S = |\overline{S}| = \sqrt{100^2 + 50^2} = 111.80 \text{ VA}$$

$$\theta = \tan^{-1}\left(\frac{Q}{P}\right) = \tan^{-1}\left(\frac{50}{100}\right) = 26.56°$$

$$pf = \cos(26.56°) = 0.894 \text{ lagging}$$

$$\overline{S} = \overline{V}\overline{I}^* \rightarrow \overline{I} = \frac{100 - j50}{240\angle 0°} = 0.466\angle - 26.56° \text{ A}$$

Example 2.4 *Complex Power* A load supplied from a 440 V source consumes 200 W at a lagging power factor of 0.9. Calculate the complex, reactive, and apparent powers drawn by this load, as well as the magnitude and phase of the current.

$$pf = \cos\theta = 0.9 \rightarrow \theta = \cos^{-1}(0.9) = 25.84°$$

$$Q = P\tan\theta = 200 \times \tan(25.84°) = 96.864 \text{ var}$$

$$\overline{S} = P + jQ = 200 + j96.864 \text{ VA}$$

$$S = |\overline{S}| = 222.22 \text{ VA}$$

$$P = VI\cos\theta \rightarrow I = \frac{200}{440 \times 0.9} = 0.505 \text{ A}$$

$$\overline{I} = I\angle - \theta° = 0.505\angle - 25.84°$$

2.11 Parallel Loads

Figure 2.12 represents schematically a set of loads supplied from a common voltage source. Let us assume for now that we can neglect the voltage drops and losses in the wires connecting these loads to the source. In Chapter 8, we will explore how we can handle this

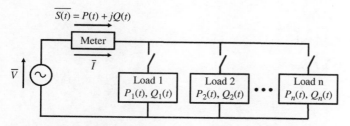

Figure 2.12 Active, reactive, and complex powers consumed by a set of loads.

more rigorously. At a given time t the total active power and reactive power required to supply these loads are given by:

$$P(t) = \sum_{i=1}^{N} P_i(t)$$

$$Q(t) = \sum_{i=1}^{N} Q_i(t) \tag{2.10}$$

where the summations are carried out over all the loads that are connected at time t. We included the time variable in these expressions to emphasize that the total reactive power load that must be supplied by the source, like the active power load, varies over time.

Reactive power is the unavoidable consequence of the fact that inductance forces the current to lag the voltage, while capacitance makes it lead the voltage. If the power factor is unity ($pf = \cos\theta = 1.0$), the current is in phase with the voltage, there is no reactive power, and all the current is put to practical use, i.e., delivering active power. While this unity power factor ideal is difficult to achieve, maintaining a power factor close to unity is good practice as we will see that excessive flows of reactive power create significant problems in the transmission and distribution networks. Since the power factor tends to be lagging because many loads are inductive, it is often corrected by adding capacitors or capacitive loads in parallel. The meters installed on the premises of smaller consumers measure only the flow of active power, and these consumers thus do not pay for reactive power. On the other hand, larger consumers are often penalized if their power factor drops below a given threshold.

Example 2.5 *Parallel Loads* Three loads are connected in parallel and supplied by an ideal 220 V voltage source:

L_1: 200 W at unity power factor
L_2: 400 W at 0.8 power factor lagging
L_3: 300 W at 0.9 power factor leading

Calculate the total active power, reactive power, complex power, apparent power and current supplied by this source, as well as the power factor of the aggregated load.

L_1: $P_1 = 200\,W$, $Q_1 = 0$

L_2: $P_2 = 400\,W$, $\theta_2 = \cos^{-1}(0.8) = 36.87°$, $Q_2 = P_2 \times \tan\theta_2 = 300\,\mathrm{var}$

L_3: $P_2 = 300\,W$, $\theta_3 = -\cos^{-1}(0.9) = -25.84°$, $Q_3 = P_3 \times \tan\theta_3 = -145.3\,\mathrm{var}$

$$P = P_1 + P_2 + P_3 = 900 \text{ W}$$

$$Q = Q_1 + Q_2 + Q_3 = 154.7 \text{ var}$$

$$\overline{S} = P + jQ = 900 + j154.7 \text{ VA}$$

$$S = |\overline{S}| = 913.2 \text{ VA}$$

$$pf = \frac{P}{S} = 0.986$$

$$I = \frac{S}{V} = 4.15 \text{ A}$$

Example 2.6 *Reactive Power Compensation* A load of 2 MW at 0.8 power factor lagging is supplied at 69 kV. Calculate the current that it draws from the source. Repeat this calculation for the case where this inductive load is compensated by a capacitive load of −1 Mvar.

Uncompensated case:

$$pf = \cos \theta = 0.8 \rightarrow \theta = \cos^{-1}(0.8) = 36.87°$$

$$Q_{unc} = P \tan \theta = 2 \times \tan(36.87°) = 1.5 \text{ Mvar}$$

$$S_{unc} = \frac{P}{\cos \theta} = 2.5 \text{ MVA}$$

$$I_{unc} = \frac{S_{unc}}{V} = 36.23 \text{ A}$$

Compensated case:

$$Q_{comp} = Q_{unc} + Q_{cap} = 1.5 - 1 = 0.5 \text{ Mvar}$$

$$S_{comp} = \sqrt{P^2 + Q_{comp}^2} = 2.06 \text{ MVA}$$

$$I_{comp} = \frac{S_{comp}}{V} = 29.88 \text{ A}$$

Figure 2.13 illustrates this reactive compensation.

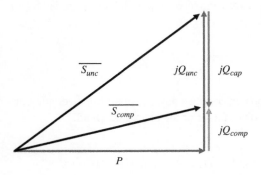

Figure 2.13 Reactive power compensation.

Reference

Auffhammer, M., Baylis, P., and Hausman, C.H. (2017). Climate change is projected to have severe impacts on the frequency and intensity of peak electricity demand across the United States. *Proceedings of the National Academy of Sciences* 114 (8): https://doi.org/10.1073/pnas .1613193114.

Further Reading

The websites of the US Energy Information Agency (www.eia.gov) and of the International Energy Agency (www.iea.org) provide lots of useful data about loads and the demand for electricity.

Problems

P2.1   Make a list of all the electrical appliances in your house, apartment, or room. Record (or estimate) the voltage and power rating for each of them.

P2.2   Using the list that you created in Problem P2.1, estimate the load profile for your house, apartment, or room on a typical day.

P2.3   Consider the simulated load profile shown in Figure P2.3. Identify or calculate the following quantities:
  • Peak load
  • Minimum load
  • Average load
  • Peak to average ratio
  • Total energy consumed
  • The capacity factor of a generator that would supply solely this load profile.
  (You may want to enter this profile in a spreadsheet as it is used in other problems).

P2.4   Many utilities and system operators publish data on their website about current and historical system load. Obtain the daily load profile for your region. If this data is not available, see www.caiso.com or www.iso-ne.com or data.nationalgrideso .com. Plot this load profile and determine the peak load, minimum load, average load, peak-to-average ratio, as well as the amount of energy consumed on a given day.

P2.5   Using the same source of data as in Problem P2.4, determine the maximum and minimum loads for the past year. On what days and at what times did they occur? Were these days particularly hot or cold?

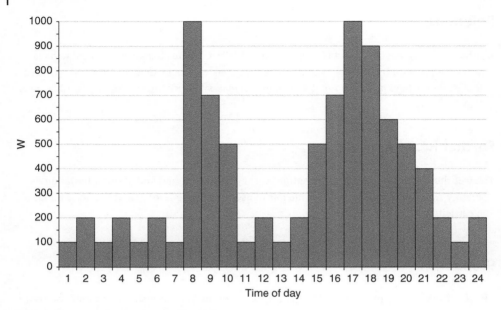

Figure P2.3 Simulated load profile of Problem P2.3.

P2.6 Many governmental and international agencies publish data on the demand for electricity. See for example www.eia.gov or www.iea.gov. Obtain data showing the evolution of the demand for electrical energy in a region or country, preferably your own. Discuss the correlation between this data and economic growth and other factors.

P2.7 Using your latest electricity bill or the website of your local electric utility, determine the rate at which residential consumers purchase electrical energy, as well as any other charge that they have to pay.

P2.8 Electric utilities often offer multiple tariffs to commercial and industrial consumers depending on their maximum demand. Using data from the website of your local electric utility, describe some of these tariffs.

P2.9 Assuming an energy rate of 0.25 $/kWh, a daily peak demand charge of 0.75 $/kW, and a daily standing charge of $0.50, calculate how much a consumer would be charged for the load profile shown in Figure P2.3.

P2.10 Some electric utilities offer their residential consumers a tariff where the rate is lower during off-peak hours. Calculate how much a consumer would be charged for the load profile shown in Figure P2.3 assuming an on-peak rate of 0.30 $/kWh, an off-peak rate of 0.20 $/kWh, a daily peak demand charge of 0.75 $/kW, and a daily standing charge of $0.50. The off-peak hours are hours 1–6.

P2.11 The owner of the house whose load profile is shown in Figure P2.3 is considering replacing a tankless water heater by a time-controlled water heater equipped with a storage tank. The operation of this new water heater would change the load profile as follows:

- Increase the energy consumption by 0.5 kWh during hours 2, 3, and 4
- Decrease the energy consumption by 0.5 kWh during hours 8, 17, and 18.

This replacement would thus not change the daily energy consumption. Calculate how much money this replacement would save each year assuming the tariff in P2.10.

P2.12 A load draws a current of 10 A from a 240 V ac source. The current lags the voltage by 20°. Calculate the corresponding active, reactive, complex, and apparent powers consumed by this load, as well as its power factor.

P2.13 A load is supplied by a 120 V ac source and draws a current of 25 A at a 0.9 pf leading. Calculate the corresponding active, reactive, complex, and apparent powers consumed by this load. Using the voltage as the reference for the angles, express the voltage and currents as phasors.

P2.14 A load of 510 kW at 0.85 pf lagging draws a current of 20 A. What is the value of the supply voltage? How much reactive power does this load consume?

P2.15 A load of 5 kW at 0.8 pf lagging is supplied at 240 V 60 Hz. Model this load using a resistance and an inductance in series. Repeat the calculations for a resistance and an inductance in parallel.

P2.16 A load of 5 MW at 0.8 pf lagging is supplied from a 13 kV 60 Hz source. Calculate the reactive compensation needed to improve the power factor to 0.85, 0.9, 0.95, and 1.0. In each case, express the value of the compensation in Mvar, Ω, and mF. Calculate the magnitude and phase of the current drawn from the source.

P2.17 An aggregated load is supplied from a 13 kV 60 Hz source and consists of the parallel combination of:

- A load of 10 MW at 0.9 pf lagging
- A load of 15 MVA at 0.85 pf lagging
- A resistance of 80 Ω in series with an inductive reactance of 10 Ω
- A resistance of 70 Ω in parallel with an inductive reactance of 200 Ω.

Calculate the reactive compensation required to achieve unity power factor. Calculate the magnitude and phase of the current drawn from the source with and without compensation.

3

Primary Energy Conversion

3.1 Overview

Depending on the primary source of energy, the process of generating electrical energy involves different conversion stages. Conventional thermal generation uses the combustion of a fuel to raise the temperature of a pressure fluid. The expansion of this fluid in a turbine produces mechanical energy, which is converted into electrical energy using a rotating synchronous generator. Wind and hydro turbines convert the energy contained in a naturally occurring fluid into mechanical energy and hence into electrical energy. Photovoltaic generation does not involve rotating machines. Instead, the solar panels produce dc electric power, which is converted into ac and injected into the grid using power electronics.

This chapter describes these various energy conversion processes and discusses their advantages and disadvantages. Chapter 4 describes how mechanical power is converted by generators into electrical power, while Chapter 5 discusses the operation and applications of power electronics converters.

3.2 Wind Generation

3.2.1 How Much Power is There in the Wind?

Consider the wind turbine shown in Figure 3.1. The turbine blades sweep an area $A = \pi r^2$. If the wind speed is v, the mass of air passing through this area in time t is:

$$m = \rho_a A v t \tag{3.1}$$

where ρ_a is the specific mass of the air in kg/m³. The kinetic energy in this mass of air is thus:

$$E_k = \frac{1}{2} m v^2 = \frac{\pi}{2} \rho_a r^2 v^3 t \tag{3.2}$$

The power that the turbine blades try to capture is then given by:

$$P_w = \frac{dE_k}{dt} = \frac{\pi}{2} \rho_a r^2 v^3 \tag{3.3}$$

Power Systems: Fundamental Concepts and the Transition to Sustainability, First Edition. Daniel S. Kirschen.
© 2024 John Wiley & Sons Ltd. Published 2024 by John Wiley & Sons Ltd.
Companion website: www.wiley.com/go/kirschen/powersystems

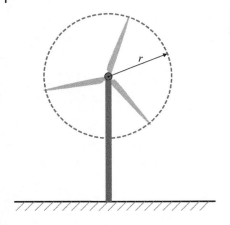

Figure 3.1 Area swept by the blades of a wind turbine.

However, it is impossible to extract all this power because this would require bringing the wind speed down to zero. It can be shown that an ideal wind turbine would not be able to extract more than 59.3% of the wind power. This limit is known as the Betz limit. In practice, the power extracted by the turbine blades is given by:

$$P_{\text{blades}} = C_p P_w \tag{3.4}$$

where C_p is the coefficient of performance of the turbine and $C_p < 0.593$.

3.2.2 How Does a Turbine Blade Extract Wind Power?

Figure 3.2 illustrates the flow of air around a cross-section of a turbine blade. If it looks like the cross-section of an airplane wing (a.k.a. an airfoil), this is no accident because the process that allows planes to fly also makes wind turbines turn. As the airfoil deflects the airflow, and thus exerts a force on the air, Newton's third law (action = reaction) implies that an equal and opposite force is applied to the airfoil. In the case of an airplane, this lift

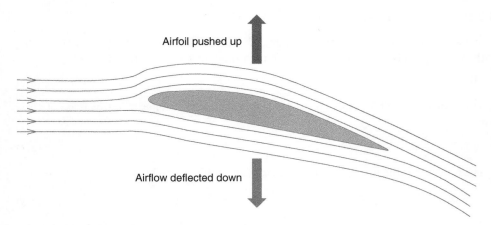

Airfoil pushed up

Airflow deflected down

Figure 3.2 Flow of air around an airfoil.

Figure 3.3 Torque production in a wind turbine.

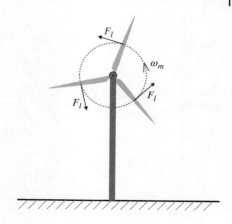

force counteracts gravity. For a wind turbine, as illustrated in Figure 3.3, this lift force F_l creates a torque on the shaft that connects the blades:

$$T_m = \frac{P_{\text{blades}}}{\omega_m} \tag{3.5}$$

where ω_m is the rotational speed of this shaft in rad/s.

This shaft is connected either directly or through a gearbox to the synchronous generator that converts the mechanical power imparted by this torque into electric power.

3.2.3 Controlling a Wind Turbine

Two main variables determine how much power a turbine extracts from the wind:

- The angle of attack, which is the angle between the direction of the wind and the chord of the airfoil (Figure 3.4). This angle can be regulated by rotating the blade at the point where it connects to the shaft of the turbine. If this angle is too big or too small, the coefficient of performance C_p decreases from its maximum value.
- The tip speed ratio:

$$\lambda = \frac{v_{\text{tip}}}{v} = \frac{\omega_m r}{v} \tag{3.6}$$

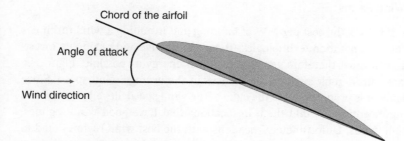

Figure 3.4 Angle of attack of an airfoil.

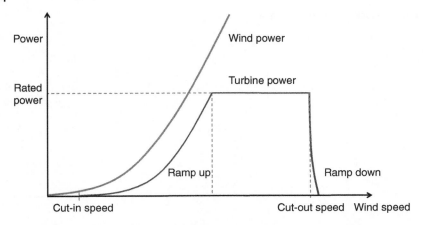

Figure 3.5 Typical power versus wind speed characteristic of a wind turbine.

where v_{tip} is the speed at which the tip of blade moves, r is the length of the blade, and v is the wind speed. This ratio can be controlled by adjusting the rotational speed of turbine ω_m. If λ is too small, too much wind passes through the turbine without affecting the blades. On the other hand, if λ is too large, the turbulence caused by the passage of a blade does not have time to dissipate before the next blade arrives. This turbulence decreases the coefficient of performance C_p.

These two variables must therefore be controlled not only to maximize the extraction of wind power, but also to operate within safe limits and avoid mechanical damage.

Figure 3.5 illustrates the power versus wind speed characteristic of a typical wind turbine. As Eq. (3.1) shows, the wind power increases with the cube of the wind speed. However, wind turbines do not generate any power until the wind speed reaches a cut-in speed. As the wind speed increases above that threshold, the angle of attack and the tip speed ratio are regulated to extract as much wind power as possible. However, once the power output of the turbine reaches its rated value, the angle of attack is reduced to avoid overloading the turbine. If the wind speed exceeds a cut-out threshold, the angle of attack of the blades is adjusted ("feathered") so that they don't generate any power and the turbine is stopped to prevent mechanical damage.

3.2.4 Locating Wind Farms

While the wind itself is free, the cost per MW of building and installing a wind turbine is high compared to other generation technologies. To be competitive and recoup its investment cost, a wind farm must therefore generate as much energy as possible. Figure 3.6 shows that there are considerable regional differences in the average wind speed. Since the available wind power is proportional to the cube of the wind speed, developers of wind farms have a strong incentive to build them in locations that have good wind regimes, i.e., consistently high winds. Unfortunately, locations with the best wind regimes tend to be in sparsely populated areas or offshore, which means that developing wind generation capacity often requires a substantial expansion of the transmission network.

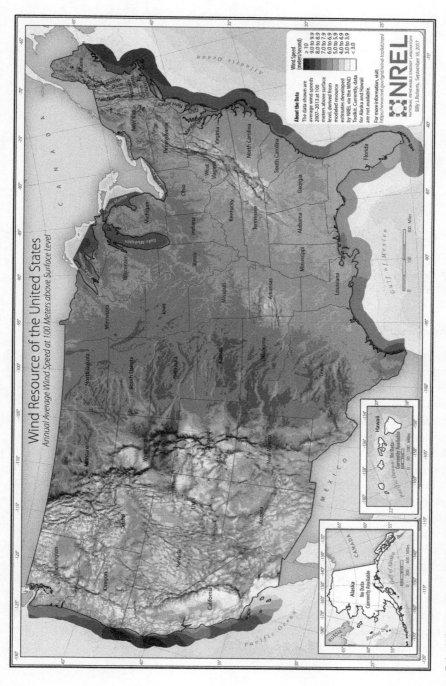

Figure 3.6 Wind resource of the United States. Source: Draxl, C., Hodge, B.M., Clifton, A. McCaa, J. (2015). *Overview and Meterological Validation of the Wild Integration National Dataset Toolkit*, Technical Report, NREL/TP-5000-61740. Golden, CO: National Renewable Energy Laboratory.

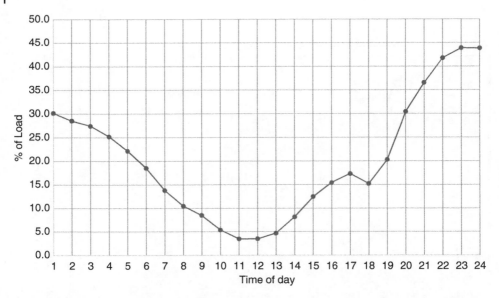

Figure 3.7 Wind generation output as a percentage of the aggregated load in the ERCOT system on November 16, 2018. Source: http://www.ercot.com.

3.2.5 Advantages and Disadvantages of Wind Generation

The main advantages of generating electrical energy from the wind are that the primary energy is free and that it does not generate greenhouse gases. The main technical disadvantage is its variability. Figure 3.7 shows the fraction of the aggregated load in the ERCOT system that was supplied by wind generation on November 16, 2018. On that admittedly rather extreme day, this fraction varied between less than 4% to almost 45%. As we will discuss in later chapters, these variations must be compensated using other forms of generation, storage, or flexibility from the load. Another issue with wind generation is its lack of controllability. Because the wind is free, wind turbines normally generate as much power as they can extract from the wind. Therefore, they can decrease their output but cannot increase it unless they operate at less than maximum output, which is economically undesirable.

The obvious economic weakness of wind generation is that it can only produce when the wind blows. Its capacity factor therefore is low. For example, Figure 3.8 shows a histogram of the wind generation output in the ERCOT system in 2018 as a percentage of the installed capacity. The overall capacity factor of the entire wind generation fleet in ERCOT for that year was 36.6%.

While wind generation is sustainable, it does have some negative environmental effects. People living near wind farms complain about their visual impact and the noise that they produce. Wind turbines also kill birds.

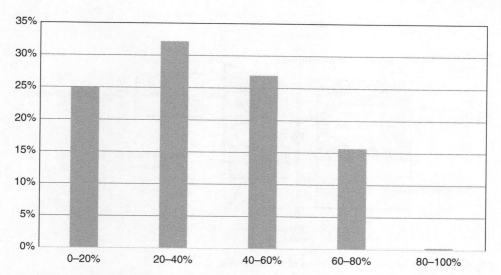

Figure 3.8 Histogram of the wind generation output in the ERCOT system in 2018 as a percentage of the installed wind generation capacity. Source: http://www.ercot.com.

3.3 Thermal Generation

3.3.1 Concept of Heat Engine

All forms of thermal electricity generation are based on the concept of heat engine. As illustrated in Figure 3.9, a heat engine consists of a heat source that delivers a quantity Q_1 of energy to a prime mover. This prime mover converts a fraction W of heat energy to mechanical energy and releases Q_2 as waste heat energy into the heat sink. French engineer Sadi Carnot demonstrated that the theoretical maximum efficiency of any heat engine is given by the following expression:

$$\eta_{\text{Carnot}} = \frac{W}{Q_1} = \frac{T_1 - T_2}{T_1} \tag{3.7}$$

where T_1 is the absolute temperature of the heat source and T_2 the absolute temperature of the heat sink. For any T_1 and T_2, the actual efficiency achieved by any practical heat engines is substantially lower than the Carnot efficiency. Nevertheless, to maximize the

Figure 3.9 Concept of heat engine.

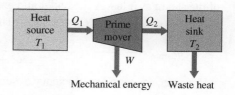

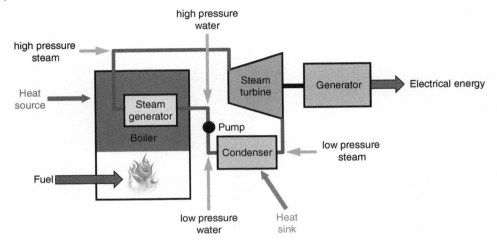

Figure 3.10 Schematic representation of a fossil-fueled steam plant.

efficiency of this energy conversion, we must maximize the difference between the temperatures of the heat source and the temperature of the heat sink. In the following sections, we discuss briefly some practical implementations of heat engines that are used for electricity generation.

3.3.2 Fossil-Fueled Steam Plants

In the basic steam cycle represented schematically in Figure 3.10, a fossil fuel such as coal or oil is burned to generate heat. This heat is used to boil high pressure water to produce high pressure steam. As this steam expands in a turbine, it produces mechanical power that drives a generator, which converts mechanical power into electrical power. The low-pressure steam coming out of the turbine must be condensed back into liquid water so that its pressure can be raised again before returning to the boiler. Without this condensation step, this heat engine would not work because raising the pressure of steam requires considerably more power than raising the pressure of water. The water is recirculated in this closed cycle because it must be demineralized to avoid damaging the steam generator and the turbine.

Example 3.1 *Carnot efficiency* In a sub-critical coal-generating plant, the temperature of the steam as it leaves the boiler and enters the turbine is 565°C. The temperature of the water as it exits the condenser is 20°C. The Carnot efficiency of this plant is:

$$\eta_{Carnot} = \frac{T_1 - T_2}{T_1} = \frac{(565 + 273.15) - (20 + 273.15)}{(565 + 273.15)} = 65\%$$

However, the practical efficiency of such a plant is typically around 39%.

Figure 3.11 illustrates the steady state relation between the electrical power produced by a steam plant and the rate at which fuel is injected in the boiler. In such an input/output

Figure 3.11 Input/output curve of a thermal generating plant.

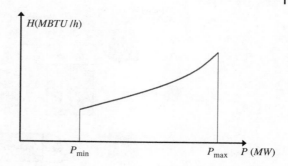

graph, the vertical axis measures the energy content of the fuel needed each hour to obtain the constant electrical power output shown on the horizontal axis. In the United States, the energy content of a fuel is usually measured in Million British Thermal Unit (MBTU)[1].

Example 3.2 *Heat rate and CO_2 emissions* According to the United States Environmental Protection Agency, the average, annual operating heat rate of U.S. coal-fired power plants is approximately 10,400 BTU/kWh. To generate a constant 500 MW, a power plant would therefore have to produce 5.2×10^9 BTU of heat every hour, which would require burning about 168,477 kg (371,429 pounds) of coal and would release 481,845 kg (1,062,360 pounds) of CO_2.

Figures 3.10 and 3.11 do not reflect the complexity of the thermodynamic system linking the injection of fuel to the production of electricity. The characteristics of this system and the need to avoid damaging its mechanical components limit the rate at which the power output of the plant can be increased or decreased. They also preclude fast or frequent startup and shutdowns. To maintain the stability of the combustion, steam plants must always generate at least a substantial fraction P_{min} of their rated capacity P_{max}.

In spite of these dynamic limitations, steam plants have the advantage of being controllable, i.e., their output can be adjusted based on the needs for electric power and they can run when needed because large quantities of fuel can be easily stored. For a long time, coal-fired and other fossil-fueled steam plants produced the bulk of electrical energy in many parts of the world. However, they are being phased out not only because they emit considerable amounts of carbon dioxide and other pollutants, but also because they are no longer competitive against other generation technologies.

3.3.3 Other Types of Steam Plants

3.3.3.1 Nuclear Power Plants

Nuclear power plants rely on essentially the same steam cycle as shown in Figure 3.10, but with the heat energy stemming from the fission of enriched uranium. Most of them are designed to operate at a constant power output. Since their fuel is relatively cheap and their operation does not produce greenhouse gases, their proponents argue that they represent an

---

1 MBTU = 10^6 BTU = 1055 MJ = 0.29307 MWh; 1 MBTU/h = 0.29307 MW.

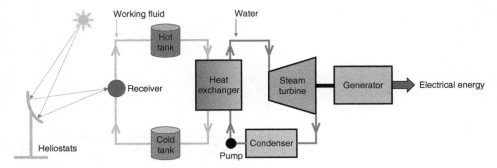

Figure 3.12 Concentrated solar power generation.

ideal way to provide the base of an aggregated load profile. However, building nuclear plants to the level of safety needed to avoid devastating accidents is proving difficult and extremely costly. Investing in nuclear power has therefore become a risky economic proposition, even if public safety concerns could be overcome.

3.3.3.2 Concentrated Solar Power Plants

In a concentrated solar power plant, solar energy is concentrated at the focal point of parabolic mirrors where it heats a working fluid, such as a molten salt, to a very high temperature. As it passes through a heat exchanger, this hot fluid provides the heat source for a conventional steam cycle. As Figure 3.12 illustrates, the hot working fluid can be stored to make it possible to continue electricity generation after the sun sets. Deployment of concentrated solar generation has so far been limited because the rapid decrease in the cost of photovoltaic panels make it a financially less attractive option.

3.3.3.3 Geothermal Plants

In some parts of the world, hot and porous rocks are located reasonably close to the surface of the earth. It is then possible to inject water at high pressure in one well and extract steam from another well. This steam can then be used either directly or indirectly to drive a turbine. Geothermal energy is sustainable and provides a useful contribution to the base load. Unfortunately, the number of locations where geothermal energy could be harnessed economically appears limited.

3.3.3.4 Combined Heat and Power Plants – Cogeneration

While there is a practical limit to the energy that a turbine can extract from steam for electricity generation, the heat energy contained in the steam that exits the turbine can be used for district heating or industrial processes. Figure 3.13 illustrates the concept of combined heat and power (CHP). Since most CHP or cogeneration schemes are designed to produce electricity as a by-product of heat rather than the other way around, electrical power is often a small fraction of the total output. Furthermore, since these systems are "heat driven," the electrical power production usually cannot be adjusted to follow the electrical load. While CHP significantly increases the overall efficiency of the energy conversion process, most schemes still rely on burning fossil fuels.

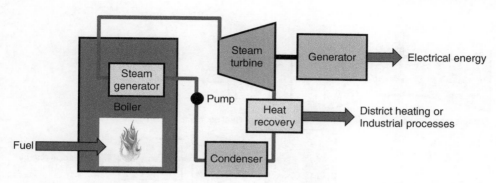

Figure 3.13 Schematic representation of a combined heat and power generating plant.

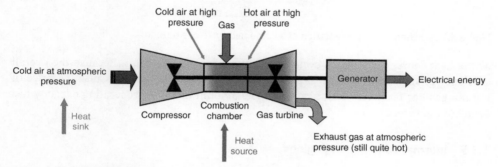

Figure 3.14 Schematic representation of an open cycle gas turbine.

3.3.4 Gas-Fired Generation

3.3.4.1 Open Cycle Gas Turbines (OCGT)

In an open cycle gas turbine (Figure 3.14), air is compressed before being injected in the combustion chamber where it is heated by the burning gas and thus gains energy. As it expands in the turbine, this hot, compressed air produces mechanical power. Since the compressor, turbine, and generator are on the same shaft, the difference between the power produced by the turbine and the power consumed by the compressor can be converted by the generator into electrical power. The exhaust contains a considerable amount of heat energy that is wasted.

Open cycle gas turbines are relatively cheap to build but expensive to operate because they are inefficient. However, because they can be started and begin injecting power in the grid in minutes and can ramp their output up and down rapidly, they are useful when power system operators must deal with a rapid increase in load or a sudden shortfall in generation.

3.3.4.2 Combined Cycle Gas Turbines (CCGT)

The development of new ceramic materials has made possible the design of turbine blades able to withstand extremely high temperatures. Increasing this temperature improves the efficiency of the gas turbine. However, this also increases the temperature of the exhaust gases. As Figure 3.15 illustrates, the remaining heat energy in these exhaust gas can serve

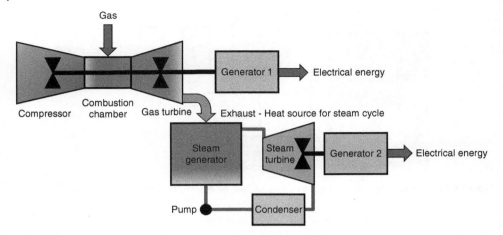

Figure 3.15 Schematic representation of a combined cycle gas turbine.

as the heat source for a steam cycle that drives a second generator. Because the overall efficiency of such combined cycle power plants can reach 63% and because the price of natural gas has been very low, a considerable amount of CCGT-generating capacity has been built in recent years.

3.3.5 Internal Combustion Engines

Generators can also be driven by internal combustion engines. However, because of the high cost of diesel fuel, such power plants only make economic sense where the required capacity is too small to justify the installation of a turbine, i.e., in small, isolated communities or for emergency backup generation.

3.4 Hydroelectric Generation

3.4.1 Impoundment Hydro Plants

All large hydro generation schemes involve impounding water in a reservoir behind a dam. A gate controls the amount of water that flows through the penstock to drive the turbine and generate electricity. As Figure 3.16 illustrates, the head is defined as the difference in elevation between the turbine and the average height of the reservoir. The potential energy per unit of volume of water stored in the reservoir is given by:

$$e = \rho_w gh \tag{3.8}$$

where ρ_w is the specific mass of water (kg/m³), g is the acceleration of gravity (m/s²), and h is the head (m).

The total potential energy stored in the reservoir is:

$$E = \rho_w ghV \tag{3.9}$$

where V is the volume of water in the reservoir (m³).

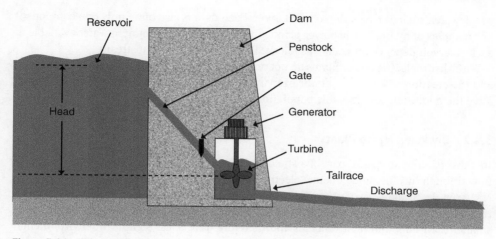

Figure 3.16 Schematic representation of an impoundment hydro plant.

Example 3.3 *Energy in a hydro reservoir* The reservoir behind the Grand Coulee dam in Washington State has a volume of 9,562,000 acre-ft (12 km³). However, only 5,185,400 acre-ft (6 km³) are considered "active," i.e., can be used for power generation. The hydraulic head is 116 m. The useable potential energy in this reservoir is thus approximately:

$$E = 1000\frac{\text{kg}}{\text{m}^3} \times 9.81\frac{\text{m}}{\text{s}^2} \times 116\,\text{m} \times 6 \times 10^9\,\text{m}^3 = 6.83 \times 10^{15}\text{J} = 1897\,\text{GWh}$$

which is about the amount of energy that the consumers of the city of Seattle use in 10 weeks.

Letting water flow through the dam at a rate of Q_w (m³/s) releases the following amount of power:

$$P_w = \rho_w g h Q_w \qquad (3.10)$$

However, due to losses in the penstock, the turbine, and the generator, the amount of electrical power generated is:

$$P = \eta \rho_w g h Q_w \qquad (3.11)$$

where the overall efficiency of the conversion process η is typically about 90%.

The power produced from a hydroelectric dam can be adjusted by opening or closing the gate to control the flow of water. However, the efficiency of hydro turbines drops sharply when the water flow is substantially below the optimal value for which the turbine was designed. To be able to adjust their power output while maximizing efficiency, hydroelectric plants therefore usually comprise several turbines that are either shut down or operated at, or close to, their rated output.

Hydro power plants are extremely flexible because they can be started rapidly, and their output ramped up or down very quickly. However, this technical flexibility is often constrained by environmental and system consideration. Extreme values of the water discharge or rapid changes in flow are harmful to the wildlife and incompatible with the other uses of the water, such as navigation or recreation. Each dam is also often only one component of a complex of hydroelectric power plants. In such cases, the amount of water available in each reservoir depends on what was released from dams located upstream.

While hydroelectric power plants do not produce greenhouse gases, the erection of a dam and the creation of a large reservoir irreparably damage the river environment and often force the relocation of vulnerable populations.

3.4.2 Diversion Hydro Plants

In a diversion hydro plant, part of a stream of water is diverted from its normal course to flow through turbines and generate electricity. Since such schemes do not store water in a reservoir, the availability of water dictates their ability to produce electric power. They are therefore often called "run of the river" plants. Diversion hydro schemes are usually much smaller than impoundment schemes and therefore have a less significant environmental impact.

3.5 Photovoltaic Generation

The amount of power available from the sun is quantified by the global horizontal solar irradiance, which is defined as the total amount of shortwave solar radiation received from above by a horizontal surface. Its units are W/m^2. Figure 3.17 shows the distribution of global horizontal solar irradiance across the United States, averaged over a year. This irradiance clearly varies widely over time depending on the season, the time of day, and the cloud cover. Roughly speaking, from a physical perspective, irradiance characterizes the number and energy of the photons hitting a certain area. Solar cells, which are the building block of photovoltaic (PV) generation, consist of the junction of p-doped and n-doped semiconductors. When photons strike this junction, they separate electrons from their atoms and create a dc current. To understand how to optimize the conversion of photons into electrical power, we need to examine the relation between the voltage and current in a PV cell as illustrated in Figure 3.18. The voltage is maximum (V_{OC}) when the cell is in open circuit, while the current is maximum (I_{SC}) when the cell is short circuited. However, the solar power extracted from the cell is maximum when the product of the voltage and current is maximum, which occurs at the knee (V^*, I^*) of the $I - V$ curve. Figure 3.19 shows how the $P - V$ characteristic of a cell varies with the solar irradiance, and in particular how the voltage corresponding to the maximum power point varies in a nonlinear fashion with the irradiance. Furthermore, since this optimal voltage also varies with the temperature of the cell, it must be adaptively adjusted to maximize power generation.

Figure 3.19 also shows that the voltage and power produced by a single cell are relatively small. Solar cells are therefore arranged in PV panels where they are connected in series to increase the overall voltage rating and in parallel to increase the current rating. In turn, parallel and series combinations of panels form a PV array.

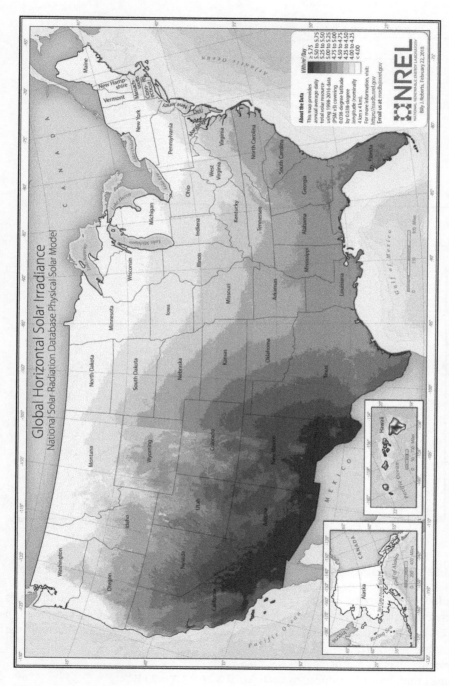

Figure 3.17 Solar resources in the United States. Source: National Renewable Energy Laboratory/https://www.nrel.gov/gis/solar-resource-maps.html.

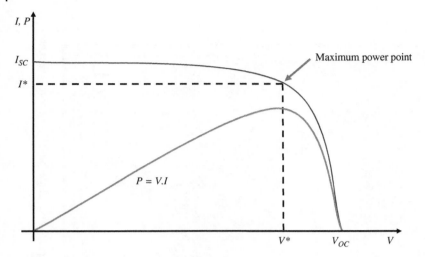

Figure 3.18 Typical *I–V* and *P–V* characteristics of a photovoltaic cell.

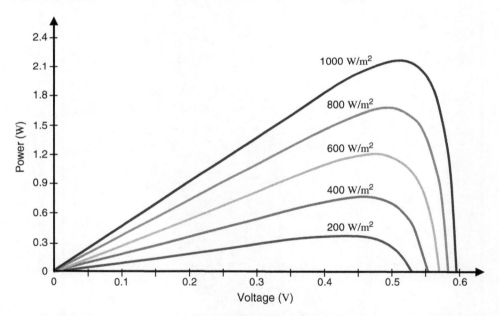

Figure 3.19 P-V characteristic of a typical photovoltaic cell as a function of the solar irradiance.

Example 3.4 *PV array* A small PV array consists of two parallel columns of four panels connected in series. Each of these panels comprises four rows of nine cells connected in series. Each cell produces a maximum of 2.5 W at 0.5 V. Calculate the voltage, current, and power rating of this array.

The voltage in each row of a panel is $9 \times 0.5 = 4.5\ V$.

The maximum power produced by each panel is $36 \times 2.5 = 90\ W$.

The current produced by each panel is then: $I_{\text{panel}} = \dfrac{90}{4.5} = 20\ A$.

Since each column of the array consists of four panels in series, the voltage rating of the array is:

$$V_{array} = V_{panel} \times 4 = 18\,V.$$

Since there are two columns in parallel, the current rating of the array is:

$$I_{array} = I_{panel} \times 2 = 40\,A$$

Finally, the power rating is given by:

$$P_{array} = V_{array} \times I_{array} = 18 \times 40 = 720\,W$$

PV panels capture a maximum of solar radiation when they are perpendicular to the direction of the sun. Maintaining this orientation as the sun moves across the sky over the course of the day requires rotating the panel around a horizontal axis to follow the elevation of the sun above the horizon, and around a vertical axis to follow its azimuth. While tracking the sun produces more energy, the cost of a dual-tracking system is often not justified. Currently, about half of the utility-scale solar installations use single-axis tracking, a small fraction uses dual-axis tracking, and the rest (as well as most residential installations) are fixed-tilt installations.

PV installations convert typically between 15% and 20% of solar irradiance into electrical energy. Greater efficiency solar cells have been developed, but their higher cost precludes for now their commercial deployment.

Photovoltaic generation has a near zero operating cost. However, on average its annual capacity factor is only about 25% in the United States. As of 2021, its construction cost (in \$/kW) is higher than for onshore wind turbines and gas-fired generation but is decreasing faster as there are more opportunities for the technology to mature.

As with wind generation, maximizing the profitability of PV generation requires producing as much electrical power as the solar irradiance allows, which limits its controllability. Another inconvenience of PV generation is that its production profile does not match the evening peak in the aggregated load profile.

While the day-to-day environmental impact of PV generation is small, the manufacture of the panels involves a significant amount of toxic chemicals.

Because PV panels produce dc, a power electronics converter is required to convert dc to ac. We discuss these converters in Chapter 5.

3.6 Storage Systems

When we charge a capacitor, we store electrical energy in an electric field. This is very convenient when we need to store a small amount of energy, but not practical for more substantial amounts because capacitors have a very low energy density. We can also store energy by injecting a current in an inductor and creating a magnetic field. However, unless the conductor is made of superconducting material, this energy dissipates quickly due to the losses. If the need arises to store electrical energy, it must be converted to potential, chemical, or mechanical energy and converted back to electrical energy when needed.

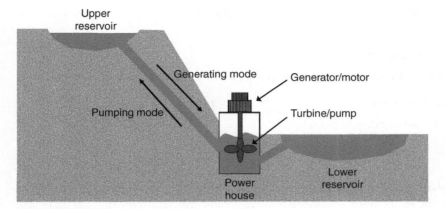

Figure 3.20 Schematic representation of a pumped hydro plant.

3.6.1 Pumped Hydro Plants

Figure 3.20 illustrates the structure and operation of a pumped hydro plant. To store energy, an electric motor extracts electrical energy from the grid and uses it to pump water from a lower reservoir into an upper reservoir. Since this reservoir is located at a higher elevation, electrical energy is converted to potential energy. When electrical energy is required, water is released from the upper reservoir, the pump reverses direction and becomes a turbine, the electric motor becomes a generator and injects power back into the grid. Pumped hydro plants can typically store enough energy to generate at rated power capacity for six to eight hours. In the current state of the technology, pumped hydro schemes can store considerably more energy and deliver more power than other forms of energy storage. However, building such plants is very costly, and the number of suitable sites where the environmental impact would be acceptable is limited.

3.6.2 Electrochemical Batteries

Advances in electrochemistry, in particular the development of the lithium-ion batteries, have made the deployment of battery energy storage a competitive solution. Batteries used to store energy in a power system are encapsulated in a battery energy storage system (BESS), whose components are illustrated in Figure 3.21. At the heart of a BESS, sits a set of battery cells arranged in series and parallel to achieve suitable voltage and current ratings. The dc terminals of this array are connected to a bidirectional power conversion system that converts dc into ac when the battery is discharging and ac into dc when the battery is charging. The ac side is single-phase for residential systems and three-phase for utility-scale systems. A battery management system (BMS) monitors voltages, temperatures, and currents in the cells to balance their loading during charging and discharging and maintain the battery within safe operating conditions. The BESS controller supervises the power conversion system in response to external charging and discharging instructions while relying on information from the BMS about the state of the battery. The amount of electrical charge that the cells can hold determines the energy rating of a battery energy storage system. The rating of the power conversion system sets the rate at which this energy

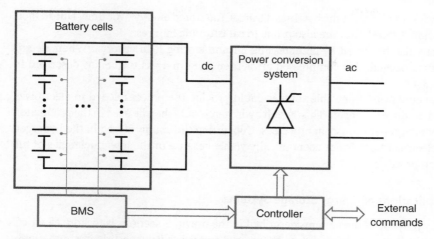

Figure 3.21 Schematic representation of a battery energy storage system.

can be injected or extracted. Dividing the energy rating of a BESS (in MWh or kWh) by its power rating (in MW or kW) gives the nominal number of hours during which the BESS is able to deliver full power. The rate at which a battery is actually charged or discharged is called the C-rate. For example, discharging a battery at a C-rate of one from a full charge means that it would be fully discharged in one hour. The state of charge (SoC) of a battery is the fraction of its energy capacity that is currently stored.

Due to losses during the electrochemical conversion process, the typical roundtrip efficiency of a BESS is about 80%. However, a fairly wide range of efficiency values have been reported in practical applications. Another significant issue is the reduction in energy capacity as the battery undergoes charge and discharge cycles. Battery degradation is a complex chemical phenomenon that depends on the frequency and depth of the discharge cycles, as well as the C-rate. Batteries used in power system applications are typically designed to have a maximum C-rate of one because charging or discharging at a faster rate increases the losses within the battery and significantly reduces its useful life. It is estimated that a battery for power system applications that undergoes one full charge/discharge cycle per day would retain only about 75% of its energy capacity after 10 years.

Example 3.5 *Battery energy storage system*   The BESS of a demonstration microgrid is rated at 1 MW, 1.38 MWh. The cells are arranged in 11 parallel racks of 9 modules in series. The voltage on the dc side varies between 1126 V and 1461 V depending on whether the battery is charging or discharging. The voltage on the ac side is 630 V. It is housed in a $23.3' \times 8' \times 9.6'$ container that weighs 55,000 pounds.

3.6.3 Other Energy Storage Technologies

Several other energy storage technologies have been proposed. Some of them have been demonstrated in pilot projects but have not yet reached commercial deployment.

- Flywheels rotating at speeds of up to 100,000 revolutions per minute (rpm) store energy in kinetic form. They have a large power rating, a long life, and require little maintenance.

However, friction losses in the bearings dissipate the stored energy over time. Due to their high rotational speed, they are also prone to catastrophic failures.

- Energy can also be stored by compressing air and storing it in caverns. When electrical energy is needed, this compressed air is then used in a gas turbine as described in Section 3.3.4.
- Surplus energy from renewable sources could provide the power needed to electrolyze water and produce hydrogen. This "green hydrogen" could then be used in the production of fertilizers or other industrial processes. Converting the energy stored in the hydrogen back to electrical form is not economically viable because of the low efficiencies of the conversion processes.

3.6.4 Applications of Energy Storage Systems

The primary purpose of the systems described in the previous sections is to store electrical energy when it is abundant, and thus cheap, and to release it when it is in short supply and thus expensive. This process is called arbitrage and is characterized by the adage "buy low, sell high." However, due to losses inherent in any energy conversion, the amount of electrical energy returned to the grid is always smaller than the amount extracted. To compensate for these losses, and for the cost of building an energy storage facility, the differences between high prices and low prices must be sufficiently high.

As we mentioned in Chapter 1, and as we will discuss in more detail in Chapter 6, the power injected into the system must be equal to the power extracted from the system. Failures of large generators occasionally cause large disturbances in this balance, while fluctuations in the load or the generation from renewable resources cause frequent but smaller perturbations. Storage systems are particularly well suited at counteracting these imbalances because they are able to adjust their output rapidly. Storage systems can therefore earn additional revenues by providing balancing services.

3.7 Choosing a Generation Technology

Historically, the choice of a technology for generating electrical energy in a given region was determined by the local availability of primary energy sources, such as hydro, coal, gas, oil, or nuclear. Government policy also influences these decisions. For example, some countries prefer to rely on their own resources rather than on imports, even though the latter might be cheaper. In recent years, many governments have also implemented policies that favor or mandate the development of electricity generation from renewable sources. If various technologies are viable, they must be compared in terms of their economics, in particular the interplay between their investment and operating costs. Let us develop a model to study this interaction. For simplicity, we will consider a single year rather than the typical lifetime of a power plant, and we will ignore the various uncertainties that decision makers must take into account.

The profit that a power plant would generate by selling an amount E of energy at an expected price π is:

$$\Gamma = \pi E - C \tag{3.12}$$

where the cost C is the sum of the operating cost of the plant and of its annuitized investment cost. If we assume that the investment cost is proportional to the capacity P of the plant, and its operating cost is proportional to the amount of energy E produced each year, we have:

$$C = c_{\text{capacity}}P + c_{\text{energy}}E \tag{3.13}$$

where c_{capacity} is expressed in \$/MW and c_{energy} in \$/MWh. The amount of energy produced each year is equal to the capacity of the plant multiplied by the expected capacity factor C_f and the number of hours in a year:

$$E = 8760 \times C_f P = C_h P \tag{3.14}$$

The expression for the profit then becomes:

$$\Gamma = \pi \, C_h P - c_{\text{capacity}}P - c_{\text{energy}}C_h P \tag{3.15}$$

If we ignore potential economies of scale, the profit per MW of installed capacity is then:

$$\frac{\Gamma}{P} = (\pi - c_{\text{energy}})C_h - c_{\text{capacity}} \tag{3.16}$$

And the energy price at which a plant breaks even is thus:

$$\pi = c_{\text{energy}} + \frac{c_{\text{capacity}}}{C_h} \tag{3.17}$$

To be competitive, a technology with a high investment cost per MW c_{capacity} must therefore have a low per MWh operating cost c_{energy} and a large capacity factor C_h.

Example 3.6 *Breakeven energy price* The table below shows rough estimates of the cost coefficients and capacity factors for a wind farm and a CCGT plant and compares the minimum average price at which these plants must sell their output to break even.

Cost, capacity factor, and breakeven energy price of a wind farm and a gas-fired power plant

	c_{energy} (\$/MWh)	c_{capacity} (\$/MW)	C_f (%)	C_h (h)	π (\$/MWh)
Wind farm	0	140,000	40	3504.0	39.95
CCGT	26.5	88,200	87	7621.2	38.07

The levelized cost of energy (LCOE) provides a more rigorous metric for comparing different generation technologies because it considers the lifetime cost of building and running a plant as well as the amount of energy that it is expected to produce over its lifetime. It is measured in \$/MWh and defined as follows:

$$\text{LCOE} = \frac{\sum_{t=0}^{N} \frac{I_t + M_t + F_t}{(1+r)^t}}{\sum_{t=0}^{N} \frac{E_t}{(1+r)^t}} \tag{3.18}$$

where N is the expected lifetime of the plant in years, I_t is the expected investment cost in year t, M_t is the expected maintenance cost in year t, F_t is the expected fuel cost in year t, E_t is the expected energy production in year t, and r is the discount rate. Future costs and energy

Table 3.1 U.S. EIA estimates of the unweighted LCOE for new generation resources entering service in 2025 (2019 dollars).

Plant type	Capacity factor (%)	LCOE ($/MWh)
Ultra-supercritical coal	85	76.44
CCGT	87	38.07
OCGT	30	66.62
Advanced nuclear	90	74.88
Geothermal	90	35.43
Biomass	83	94.83
Wind, onshore	40	39.95
Wind, offshore	44	122.25
Solar photovoltaic	29	33.12
Hydroelectric	59	52.79

Source: Adapted from www.eia.gov/aeo/pdf/electricity_generation.pdf.

productions are decreased by $(1 + r)^{-t}$ discount factors to reflect that money and energy t years in the future are perceived as having less value than they do at the present time ($t = 0$). Table 3.1 shows U.S. Energy Information Agency's estimates of LCOE for different types of generation technologies. Producing these estimates requires a substantial number of assumptions beyond what is included in the above formula. Furthermore, these values change from year to year as technologies evolve and market conditions change. Students are encouraged to compare these estimates with the latest numbers they are able to find.

Further Reading

Masters, G.M. (ed.) (2013). *Renewable and Efficient Electric Power Systems*, 2e. Wiley – IEEE Press.

Jenkins, N. and Ekanayake, J. (2017). *Renewable Energy Engineering*. Cambridge University Press.

Problems

P3.1 Many utilities and system operators publish data on their website about wind generation. Obtain such data for your region for the most recent year. If such data is not available, see www.caiso.com or www.ercot.com. Based on this data, determine:
- The total installed wind generation capacity
- The total amount of electrical energy produced from the wind
- The fraction of the total energy consumed that was produced from the wind
- The maximum absolute wind energy production and the day and time when it happened

- The maximum fraction of the load produced from the wind and the day and time when it happened
- The largest hour-to-hour increase in wind generation
- The largest hour-to-hour decrease in wind generation
- The largest intraday difference between maximum and minimum wind generation
- The largest intraday difference between maximum and minimum wind generation as a fraction of the load
- The capacity factor of wind generation for the whole year
- The day with the highest wind generation capacity factor.

P3.2   Using the same data as in Problem P3.1, plot on the same diagram:
- The hourly load profile on the day with the largest difference between maximum and minimum wind generation
- The hourly wind generation profile on that day.

P3.3   Using the same data as in Problem P3.1, plot:
- The capacity factor of wind generation for each month of the year
- The total energy consumed during each month of the year
- The fraction of this energy produced from wind generation during each month of the year.

Comment on these results.

P3.4   Calculate the Carnot efficiency of a supercritical coal-fired steam plant that heats the steam to 600°C and cools the water to 20°C.

P3.5   Calculate the Carnot efficiency of a gas turbine that takes in air at 20°C and heats it to 1360°C.

P3.6   A 300 MW coal-fired power plant has a 65% capacity factor. Calculate the amount of coal it burns in a year and the amount of CO_2 it releases in the atmosphere assuming a heat rate of 10,400 BTU/kWh.

P3.7   Is there hydro generation in your region? Find an example and determine whether it is impoundment, diversion, or pumped hydro. What is the volume of the reservoir? How much energy can it store? How much power can it produce? How much electrical energy did it produce last year? What is its capacity factor?

P3.8   Hoover Dam in the State of Arizona has a maximum active reservoir capacity of 19.554 km³ and a maximum head of 180 m. What is the maximum amount of energy that it can store?

P3.9   Assuming that Hoover Dam is operating at an average head of 150 m and is operating at its full installed capacity of 2080 MW, at what rate is water flowing through the dam?

P3.10 Compare the power available from a hydro generation site with a head of 200 m and a flow of 36 m³/s with that available from a site with a head of 10 m and a flow of 728 m³/s. Assume that the overall efficiencies at the two sites are identical.

P3.11 A hydro power plant is supplied from a reservoir with a surface of 5 km². When the reservoir is full, the plant has a head of 20 m. The plant has a total efficiency of 72%.
- Calculate the amount of electrical energy produced by this plant when enough water is released to lower the level of the reservoir from its maximum to 8 m below this maximum.
- What is the power produced by this plant when the reservoir is full, and water is released at a rate of 50 m³/s?

P3.12 A PV panel consists of 60 cells arranged in four parallel columns with 15 series cells in each column. Assuming that for given irradiance and temperature conditions, each cell has the idealized IV curve shown below, what is the maximum power that this panel can produce for these conditions?

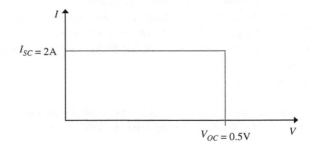

P3.13 Some utilities and system operators publish data on their website about solar generation. Obtain such data for your region for the most recent year. If such data is not available, see www.caiso.com or www.pjm.com. Based on this data, determine:
- The total installed solar generation capacity
- The total amount of electrical energy produced from the sun
- The fraction of the total energy consumed that was produced from the sun
- The maximum absolute solar energy production and the day and time when it happened
- The maximum fraction of the load produced from the sun and the day and time when it happened
- The largest hour-to-hour increase in solar generation
- The largest hour-to-hour decrease in solar generation
- The largest intraday difference between maximum and minimum solar generation
- The largest intraday difference between maximum and minimum solar generation as a fraction of the load
- The capacity factor of solar generation for the whole year
- The day with the highest solar generation capacity factor.

P3.14   Using the same data as in Problem P3.13, plot on the same diagram:
The hourly load profile on the day with the largest difference between maximum and minimum solar generation
The hourly solar generation profile on that day.

P3.15   Using the same data as in Problem P3.13, plot:
The capacity factor of solar generation for each month of the year
The total energy consumed during each month of the year
The fraction of this energy produced from solar generation during each month of the year.
Comment on these results.

P3.16   The difference in elevation between the levels upper and lower reservoirs of a pumped hydro plant is a constant 100 m. Assuming that the roundtrip efficiency of the plant is 75%, calculate the power it would re-inject in the system after pumping 4,000,000 m^3 if this water were released at a constant rate over four hours.

P3.17   The owner of a 10 MW/80 MWh battery energy storage system has paid $640 to raise its state of charge from 10% to 90%. At what price ($/MWh) should it sell this energy back to break even? Assume that the roundtrip efficiency of the storage system is 80%.

P3.18   What policy decisions have influenced or are currently influencing the choice of generation technologies in your region?

P3.19   Plot the breakeven price of energy for the wind farm and the CCGT plant of Example 3.6 as a function of their capacity factor. Discuss how these results vary as you change the assumptions about the investment and operation costs.

4

Electromechanical Power Conversion

4.1 Overview

In Chapter 3, we saw how primary energy sources, such as wind, hydro, and fossil fuels, are converted into mechanical power. In this chapter we discuss how this mechanical power is converted into ac electrical power. Since all large generators produce a three-phase ac voltage, we also introduce the basic concepts of three-phase systems, as well as the per-unit normalization. Using a simple model, we then study the operation of generators both in isolated mode and as part of a large system.

4.2 Structure of a Generator

Figure 4.1 shows a schematic axial view of a three-phase synchronous generator. Such a generator consists of a stator and a rotor. The rotor is connected to the shaft of the prime mover (i.e., the wind, steam, hydro, or gas turbine) and rotates inside the stator. Three stator windings (a–a', b–b', c–c') are arranged in slots on the inside of the stator. In this figure, each of these windings is represented by a single loop to highlight the fact that the axes of these windings are separated by 120°. In practice, each winding consists of multiple loops spread around the circumference of the stator.

A single winding, called the field or excitation winding, is wound around the rotor. An external dc source causes a dc current I_f to flow through this rotor winding. This current creates a magnetic flux Φ of constant magnitude aligned with the axis of the rotor. Since this magnetic flux crosses the airgap between the rotor and the stator, it links the stator windings. If the conductors that comprise each stator winding are properly spaced around the circumference of the stator, the flux linking each of them is a sinusoidal function of the relative positions of the rotor and stator winding axes:

$$\lambda_{a-a'} = \mathcal{N} \, \Phi \, \cos \theta$$
$$\lambda_{b-b'} = \mathcal{N} \Phi \, \cos \left(\theta - 120° \right)$$
$$\lambda_{c-c'} = \mathcal{N} \Phi \, \cos \left(\theta + 120° \right) \tag{4.1}$$

where $\mathcal{N}$ is the effective number of turns of the stator windings and θ is the angle between the rotor and the a–a' axis.

Power Systems: Fundamental Concepts and the Transition to Sustainability, First Edition. Daniel S. Kirschen.
© 2024 John Wiley & Sons Ltd. Published 2024 by John Wiley & Sons Ltd.
Companion website: www.wiley.com/go/kirschen/powersystems

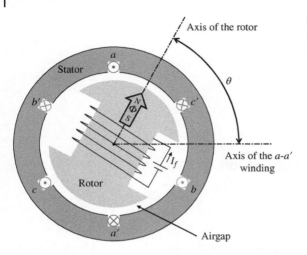

If, under the action of the prime mover, the rotor rotates at a constant speed ω, we have:

$$\theta = \omega t + \theta_0 \tag{4.2}$$

The magnetic fluxes linking the stator winding are thus a function of time and, according to Lenz's law, induce voltages in these windings:

$$e_{a-a'}(t) = \frac{d\lambda_{a-a'}(t)}{dt} = -\mathcal{N}\,\Phi\omega\,\sin(\omega t + \theta_0)$$

$$e_{b-b'}(t) = \frac{d\lambda_{b-b'}(t)}{dt} = -\mathcal{N}\,\Phi\omega\,\sin(\omega t + \theta_0 - 120°) \tag{4.3}$$

$$e_{c-c'}(t) = \frac{d\lambda_{c-c'}(t)}{dt} = -\mathcal{N}\,\Phi\omega\,\sin(\omega t + \theta_0 + 120°)$$

Figure 4.2 illustrates these voltages as one would observe them using an oscilloscope connected to the terminals of the a–a', b–b', and c–c' windings. These three voltages:

- Have the same magnitude. We can control this magnitude by adjusting the flux Φ using the field current I_f.
- Are exactly 120° out of phase in time because the axes of the three windings are shifted by exactly 120° in space.
- Have the same angular frequency ω, which is determined by the mechanical speed of the prime mover.

The rotor shown in Figure 4.1 has one pole pair, i.e., one North magnetic pole and one South magnetic pole. Each mechanical revolution of the rotor will therefore induce one electrical cycle of the voltage in each stator winding. If we wish to induce voltages at 60 Hz, the rotor must therefore perform 60 revolutions per second, or 3600 revolutions per minute (rpm). Similarly, a 50 Hz voltage would require 3000 rpm.

Rotors can be built with two or more pole pairs arranged symmetrically around its circumference. If a rotor has two pole pairs, the North magnetic pole and the South magnetic pole cross the axis of each stator winding twice during each revolution of the rotor. The induced

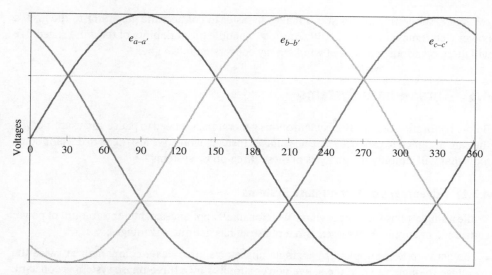

Figure 4.2 Voltages induced in the stator windings of a synchronous generator over one revolution of the rotor.

voltages therefore have twice the frequency they would have if the rotor had a single pole pair. In general, we can therefore write:

$$\omega = 2\pi f = \frac{p}{2}\omega_m \tag{4.4}$$

where f and ω are the frequency and angular frequency of the induced voltages, p is the number of magnetic poles of the rotor, and ω_m is the mechanical angular frequency of the rotor.

Since we have the following relation between the mechanical angular frequency and the rotational speed N in rpm:

$$\omega_m = 2\pi \frac{N}{60} \tag{4.5}$$

We can rewrite (4.4) as follows:

$$f = \frac{p}{120}N \tag{4.6}$$

To produce voltages at 60 Hz, a generator with a four-pole (two pole pairs) rotor therefore needs to rotate at 1800 rpm. Generators driven by gas or steam turbines tend to have one or two pole pairs because these turbines are more efficient at high speeds. Generators in hydroelectric power plants can have up to 100 poles because hydro turbines are more efficient at low speeds.

To efficiently extract power from the wind, turbines typically rotate at speeds between 5 and 20 rpm. As an example, let us suppose that a wind turbine rotates at 12 rpm. If this turbine were to drive a synchronous generator connected directly to a 60 Hz grid, (4.6) suggests that this generator would have to have 600 poles. Such a generator would be extremely heavy and would not fit in the nacelle of a wind turbine. A mechanical gearbox is therefore often used to transform the low speed of the turbine into a higher speed for the rotor of the generator. While this higher speed makes it easier to connect the generator to the grid, a power electronics converter is still required to handle the variability of the wind speed. We will discuss various schemes of wind energy conversion in Chapter 5.

4.3 Three-Phase Systems

Before continuing our study of synchronous generators, we need to pause and examine why they are designed to produce three ac voltages instead of one. We also need to develop some tools that will simplify the analysis of these three-phase systems.

4.3.1 Advantages of Three-Phase Systems

While a single-phase ac supply works well for small appliances, when the amount of power consumed or supplied increases, three-phase ac has definite advantages:

- The active power supplied by a single-phase ac source has a component at twice the voltage frequency, while the active power supplied by a three-phase system is constant. Three-phase ac motors thus run more smoothly than single-phase motors.
- It can be shown that three-phase generators, motors, and transformers have a better power-to-weight ratio than the single-phase equivalents and are thus cheaper.
- Similarly, three-phase transmission and distribution lines require less material than single-phase lines of equal rating.
- Unlike single-phase motors, three-phase motors are naturally self-starting.
- Rectification of a three-phase supply using power electronics yields a smoother dc voltage than rectification of a single-phase supply.
- A single-phase ac supply can be obtained easily from a three-phase supply, while the opposite is not practical.

4.3.2 Three-Phase Sources

Since the generator of Figure 4.1 has three stator windings, we have access to six endpoints or terminals, which we will label a, a', b, b', c, c'. However, connecting this generator to a power system or a three-phase load can be done with only three or four wires. Using fewer wires represents a considerable savings and does not involve any loss of functionality.

Let us connect together terminals a', b', and c', and let us call this point of connection n (for neutral) as shown in Figure 4.3a. Such a three-phase source is then said to be *Y-connected*. Since the voltages induced in the three stator windings have now a common reference point, we can represent them as shown in Figure 4.3b, where $\overline{V_{xy}}$ is the phasor denoting the voltage from point y to point x. These voltages are called *line-to-neutral voltages* and can be expressed as follows if we use the voltage between points n and a as the reference for the angles:

$$\overline{V_{an}} = V\angle 0°$$
$$\overline{V_{bn}} = V\angle -120° \qquad\qquad (4.7)$$
$$\overline{V_{cn}} = V\angle +120°$$

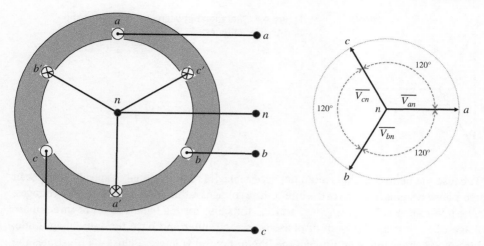

Figure 4.3 (a) Y-connection of a three-phase generator; (b) corresponding line-to-neutral voltages.

where V is the RMS value of these voltages. A set of three-phase phasors that have equal magnitude and whose phases are separated by 120° is said to be balanced.

Let us calculate the sum of these three voltages:

$$
\begin{aligned}
\overline{V_{an}} + \overline{V_{bn}} + \overline{V_{cn}} &= V\angle 0° + V\angle -120° + V\angle 120° \\
&= V + V\cos(-120°) + jV\sin(-120°) + V\cos(120°) + jV\sin(120°) \\
&= V[1 + \cos(-120°) + \cos(120°)] \\
&= 0
\end{aligned}
\tag{4.8}
$$

This is an important result that is applicable to the sum of any three balanced voltage or current phasors.

The *line-to-line voltages* are the voltages measured between the terminals a, b, and c. We can calculate their values from the line-to-neutral voltages as follows:

$$
\begin{aligned}
\overline{V_{ab}} = \overline{V_{nb}} + \overline{V_{an}} &= \overline{V_{an}} - \overline{V_{bn}} \\
&= V\angle 0° - V\angle -120° \\
&= \sqrt{3}V\angle 30°
\end{aligned}
\tag{4.9}
$$

Similarly:

$$
\begin{aligned}
\overline{V_{ca}} = \overline{V_{na}} + \overline{V_{cn}} &= \overline{V_{cn}} - \overline{V_{an}} \\
&= V\angle 120° - V\angle 0° \\
&= \sqrt{3}V\angle 150°
\end{aligned}
\tag{4.10}
$$

And:

$$
\begin{aligned}
\overline{V_{bc}} = \overline{V_{nc}} + \overline{V_{bn}} &= \overline{V_{bn}} - \overline{V_{cn}} \\
&= V\angle -120° - V\angle +120° \\
&= \sqrt{3}V\angle 270° = \sqrt{3}V\angle -90°
\end{aligned}
\tag{4.11}
$$

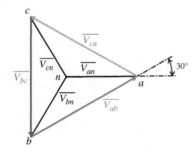

Figure 4.4 Relation between line-to-neutral and line-to-line voltages.

Because the neutral point is often not accessible, it is conventional to specify voltages in three-phase systems in terms of the line-to-line voltages rather than line-to-neutral voltages.

The line-to-line voltages $\overline{V_{ab}}$, $\overline{V_{ca}}$, and $\overline{V_{bc}}$ thus also form a balanced set of three-phase voltages, i.e., they are 120° out of phase with each other and have the same magnitude, which is $\sqrt{3}$ times the magnitude of the line-to-neutral voltages. Figure 4.4 illustrates the relation between the line-to-neutral and the line-to-line voltages. In line with (4.8), and as illustrated by the fact that they form a closed triangle, the sum of these balanced line-to-line voltage phasors is equal to zero.

Instead of connecting one end of each winding to a common point, we can connect them in a loop or Δ (delta), i.e., connect c' to a, b' to c, and a' to b, as shown in Figure 4.5a. Such a source is then said to be *Δ-connected*. In this case, since there are only three electrically distinct terminals, we have access only to the line-to-line voltages. Taking the voltage $\overline{V_{ab}}$ as a reference for the angles, we have:

$$\overline{V_{ab}} = V\angle 0°$$
$$\overline{V_{ca}} = V\angle +120° \tag{4.12}$$
$$\overline{V_{bc}} = V\angle -120°$$

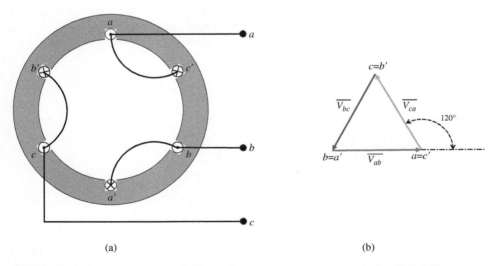

(a)   (b)

Figure 4.5 (a) Delta-connection of a three-phase generator; (b) corresponding line-to-line voltages.

As shown in Figure 4.5b, these phasors form an equilateral triangle similar to the one in Figure 4.4.

4.3.3 Three-Phase Loads

A three-phase load consists of three impedances that can be Y-connected, as shown in Figure 4.6, or Δ-connected, as shown in Figure 4.7. When these impedances are equal a three-phase load is said to be balanced.

Motors and other large loads are generally three-phase and balanced. On the other hand, most of the loads that we encounter in our daily lives are single-phase. The loading on a three-phase distribution network can therefore be imbalanced because different single-phase loads are connected to each phase. However, if we connect a sufficiently large number of single-phase loads and spread them between the three phases, the differences even themselves out and the aggregated load can be treated as balanced. In the remainder of this textbook, we will therefore assume that the three phases are balanced except when we consider faults.

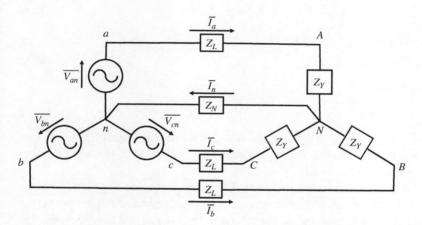

Figure 4.6 Y-connected load supplied from a Y-connected source in a four-wire system.

Figure 4.7 Delta-connected load.

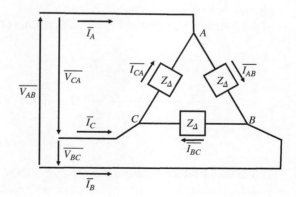

4.3.3.1 Y-Connected Loads

Let us first consider, as Figure 4.6 illustrates, the case of a balanced Y-connected load supplied from a Y-connected source. Each terminal of the source is connected to a terminal of the load (*a* to *A*, *b* to *B*, *c* to *C*) through a wire of impedance Z_L. The neutral point n of the source is connected to the neutral point *N* of the load through a wire of impedance Z_N.

Applying Kirchhoff's current law (KCL) at node *n* or *N*, we get:

$$\overline{I_n} = \overline{I_a} + \overline{I_b} + \overline{I_c} \tag{4.13}$$

Applying Kirchhoff's voltage law (KVL) around the loops *a-A-N-n-a*, *b-B-N-n-b*, *c-C-N-n-c*, we have:

$$\overline{V_{an}} = \overline{I_a}(Z_Y + Z_L) + \overline{I_n}Z_N$$
$$\overline{V_{bn}} = \overline{I_b}(Z_Y + Z_L) + \overline{I_n}Z_N \tag{4.14}$$
$$\overline{V_{cn}} = \overline{I_c}(Z_Y + Z_L) + \overline{I_n}Z_N$$

Adding these three equations, we get:

$$\overline{V_{an}} + \overline{V_{bn}} + \overline{V_{cn}} = (\overline{I_a} + \overline{I_b} + \overline{I_c})(Z_Y + Z_L) + 3\overline{I_n}Z_N \tag{4.15}$$

Combining (4.15), (4.8), (4.13) leads to:

$$0 = (\overline{I_a} + \overline{I_b} + \overline{I_c})(Z_Y + Z_L + 3Z_N) \tag{4.16}$$

which implies that:

$$\overline{I_a} + \overline{I_b} + \overline{I_c} = \overline{I_n} = 0 \tag{4.17}$$

The fact that $\overline{I_n} = 0$ implies that the voltages at nodes *N* and *n* are equal. These voltages would remain equal even if we removed the wire connecting *N* and *n*. However, if either the source or the load is unbalanced $\overline{I_n} \neq 0$ and $\overline{V_n} \neq \overline{V_N}$.

Some neutral points on the source or the load side are connected to ground or earth ("grounded"). However, one should never assume that an ungrounded neutral point is at ground potential because imbalances or harmonics in the system can raise its voltage to dangerous levels. Keeping the fourth wire alleviates but does not solve the problems caused by imbalances and harmonics. We will discuss techniques to analyze unbalanced power systems when we study faults.

If the system is balanced and $\overline{I_n} = 0$, we get from (4.14):

$$\overline{I_a} = \frac{\overline{V_{an}}}{Z_Y + Z_L} = \frac{V\angle 0°}{|Z_Y + Z_L|\angle\theta} = I\angle(-\theta)$$

$$\overline{I_b} = \frac{\overline{V_{bn}}}{Z_Y + Z_L} = \frac{V\angle - 120°}{|Z_Y + Z_L|\angle\theta} = I\angle(-120° - \theta) \tag{4.18}$$

$$\overline{I_c} = \frac{\overline{V_{cn}}}{Z_Y + Z_L} = \frac{V\angle + 120°}{|Z_Y + Z_L|\angle\theta} = I\angle(+120° - \theta)$$

A balanced load thus draws a balanced set of three-phase currents, i.e., three currents of equal magnitude that are 120° out of phase with each other.

Example 4.1 *Y-connected three-phase load* A balanced three-phase Y-connected load is supplied from a balanced three-phase voltage source with a line-to-line voltage of 480 V. The impedance of each phase of the load is $Z_Y = 9 + j4 \, \Omega$ and the impedance of each line connecting the source to the load is $Z_L = R_L + jX_L = 1 + j2 \, \Omega$.

Let us choose the line-to-line voltage between phases a and b as the reference for the angles:

$$\overline{V_{ab}} = 480\angle 0°\text{V}$$

The line-to-neutral voltage of phase a is then:

$$\overline{V_{an}} = \frac{\overline{V_{ab}}}{\sqrt{3}}\angle - 30° = 277.13\angle - 30°\text{V}$$

The current in phase a of the load is:

$$\overline{I_a} = \frac{\overline{V_{an}}}{Z_Y + Z_L} = \frac{277.13\angle - 30°}{10 + j6} = 23.764\angle - 60.96°\text{A}$$

The voltages across and the currents through the other two phases have the same magnitudes but shifted by $\pm 120°$:

$$\overline{V_{bn}} = 277.13\angle - 150° \qquad \overline{V_{cn}} = 277.13\angle + 90°$$
$$\overline{I_b} = 23.764\angle - 180.96° \qquad \overline{I_c} = 23.764\angle + 59.04°$$

The line-to-neutral voltage on phase A of the load is:

$$\overline{V_{AN}} = \overline{V_{an}} - Z_L\overline{I_a} = 277.13\angle - 30° - 23.764\angle - 60.96° \times (1 + j2) = 234.05\angle - 37.0°$$

The line-to-neutral voltages on phases B and C of the load are shifted from the voltage on phase A by 120°:

$$\overline{V_{BN}} = 234.05\angle - 157.0°$$
$$\overline{V_{CN}} = 234.05\angle 83.0°$$

The complex power supplied by phase a of the source is:

$$\overline{S_a} = \overline{V_{an}}\,\overline{I_a}^* = 277.13\angle - 30° \times 23.764\angle + 60.96° = 5647 + j3388 \text{ VA}$$

Hence:

$$P_a = 5.647 \text{ kW}$$
$$Q_a = 3.388 \text{ kvar}$$

The total power supplied by the source is three times the power supplied by one phase:

$$P_{\text{Source}} = 16.941 \text{ kW}$$
$$Q_{\text{Source}} = 10.164 \text{ kvar}$$

The complex power consumed in phase A of the load is:

$$\overline{S_A} = \overline{V_{AN}}\,\overline{I_a}^* = 234.05\angle - 37.0° \times 23.764\angle + 60.96° = 5083 + j2258 \text{ VA}$$

Hence:

$$P_A = 5.083 \text{ kW}$$

$$Q_A = 2.258 \text{ kvar}$$

The total power consumed by the load is three times the power consumed by each phase:

$$P_{\text{Load}} = 15.248 \text{ kW}$$

$$Q_{\text{Load}} = 6.775 \text{ kvar}$$

The total active and reactive losses in the lines connecting the source to the load are:

$$P_{\text{losses}} = P_{\text{Source}} - P_{\text{Load}} = 1.693 \text{ kW}$$

$$Q_{\text{losses}} = Q_{\text{Source}} - Q_{\text{Load}} = 3.389 \text{ kvar}$$

4.3.3.2 Δ-Connected Loads

Loads can also be connected in Δ, as shown in Figure 4.7. In this case, the line-to-line voltage is applied directly across each phase of the load. Taking the voltage between phases A and B as reference for the angles, we have:

$$\overline{I_{AB}} = \frac{\overline{V_{AB}}}{Z_\Delta} = \frac{V\angle 0°}{|Z_\Delta|\angle\theta} = I\angle(-\theta)$$

$$\overline{I_{BC}} = \frac{\overline{V_{BC}}}{Z_\Delta} = \frac{V\angle -120°}{|Z_\Delta|\angle\theta} = I\angle(-120° - \theta) \tag{4.19}$$

$$\overline{I_{CA}} = \frac{\overline{V_{CA}}}{Z_\Delta} = \frac{V\angle +120°}{|Z_\Delta|\angle\theta} = I\angle(+120° - \theta)$$

Applying KCL at nodes A, B, and C we have:

$$\overline{I_A} = \overline{I_{AB}} - \overline{I_{CA}} = I\angle(-\theta) - I\angle(+120° - \theta) = \sqrt{3}I\angle(-\theta - 30°)$$

$$\overline{I_B} = \overline{I_{BC}} - \overline{I_{AB}} = I\angle(-120° - \theta) - I\angle(-\theta) = \sqrt{3}I\angle(-\theta - 150°) \tag{4.20}$$

$$\overline{I_C} = \overline{I_{CA}} - \overline{I_{BC}} = I\angle(+120° - \theta) - I\angle(-120° - \theta) = \sqrt{3}I\angle(-\theta + 90°)$$

A Δ-connected load therefore also draws a balanced set of three-phase currents, whose magnitude is $\sqrt{3}$ times larger than the current flowing in each phase of the load and lag these phase currents by 30°.

Example 4.2 *Δ-connected three-phase load* A balanced three-phase Δ-connected load is supplied directly from a balanced three-phase voltage source with a line-to-line voltage of 480 V. The impedance of each phase of the load is $Z_\Delta = 10 + j6 \ \Omega$.

Let us choose the line-to-line voltage between phases a and b as the reference for the angles:

$$\overline{V_{AB}} = 480\angle 0° \text{ V}$$

The current in phase A–B of the load is then:

$$\overline{I_{AB}} = \frac{\overline{V_{AB}}}{Z_\Delta} = \frac{480\angle 0°}{10 + j6} = 41.16\angle -30.96° \text{A}$$

The current in the line connecting the source to point A of the load is then:

$$\overline{I_A} = \sqrt{3}\,\overline{I_{AB}}\angle(-30°) = 71.29\angle-60.96°\,\text{A}$$

The complex power consumed in phase AB of the load is:

$$\overline{S_{AB}} = \overline{V_{AB}}\,\overline{I_{AB}}^* = 480\angle 0° \times 41.16\angle +30.96° = 16.942 + j10.163\text{kVA}$$

$$P_{AB} = 16.942\,\text{kW}$$

$$Q_{AB} = 10.163\,\text{kvar}$$

The total power consumed by this load is three times the power consumed in each phase:

$$P = 50.826\,\text{kW}$$

$$Q = 30.489\,\text{kvar}$$

4.3.3.3 Δ-Y Equivalence

Let us consider the balanced Δ-connected load shown in Figure 4.8a, where each branch has an impedance Z_Δ. To facilitate the analysis of three-phase systems, we would like to replace it by the equivalent Y-connected load shown in Figure 4.8b. These two loads are equivalent if the impedances that appears between any two terminals are equal on both sides. Since the impedances in all phases are equal, we can choose without loss of generality to leave terminal C open in both configurations and compare the impedances between phases A and B:

For the Δ-connected load: $Z_{AB} = Z_\Delta \parallel 2Z_\Delta = \frac{2}{3}Z_\Delta$
For the Y-connected load: $Z_{AB} = Z_Y + Z_Y = 2Z_Y$
Equivalency thus requires that:

$$Z_Y = \frac{1}{3}Z_\Delta \tag{4.21}$$

Example 4.3 *Parallel Y- and Δ-connected three-phase loads* The balanced Y-connected and Δ-connected loads of Figure 4.9 are connected in parallel and are supplied from a balanced three-phase source with a line-to-line voltage $V_{LL} = 480$ V. The values of the impedances are as follows: $Z_L = 0.2 + j1\Omega$, $Z_Y = 10 + j2\Omega$, $Z_\Delta = 12 + j3\Omega$.

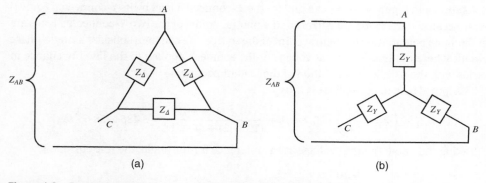

(a)   (b)

Figure 4.8 Equivalence between a Δ-connected and a Y-connected load.

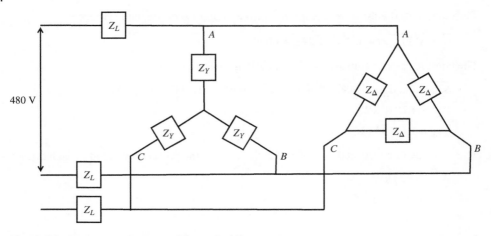

Figure 4.9 Three-phase system of Example 4.3.

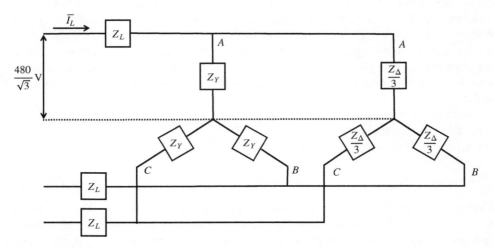

Figure 4.10 Three-phase system of Example 4.3 after Δ-Y conversion.

As shown in Figure 4.10, we can replace the Δ-connected load by its Y-connected equivalent. Because these loads are balanced, the neutral points of the two Y-connected loads are at the same potential as the neutral point of the source. We can then consider a single-phase circuit where the line-to-neutral voltage of the source is applied to the line impedance in series with the parallel combination of the load impedances.

The impedance in each phase is thus:

$$Z = Z_L + \left(Z_Y \parallel \frac{Z_\Delta}{3} \right) = 0.2 + j1 + \frac{(10 + j2)(4 + j1)}{(10 + j2) + (4 + j1)} = 3.486 \angle 28.68° \, \Omega$$

Taking the line-to-neutral voltage as the reference for the angle, the line current is:

$$\overline{I}_L = \frac{\overline{V_{LN}}}{Z} = \frac{\frac{480}{\sqrt{3}} \angle 0°}{3.486 \angle 28.68°} = 79.49 \angle -28.68° \, A$$

The complex power supplied by each phase of the source is:

$$S_{1\phi} = \overline{V_{LN}}\, \overline{I_L^*} = 19.326 + j10.572\,\text{kVA}$$

The total power supplied by the source is three times the power supplied in each phase:

$$P_{3\phi} = 3 \times P_{1\phi} = 57.979\,\text{kW}$$

$$Q_{3\phi} = 3 \times Q_{1\phi} = 31.717\,\text{kvar}$$

4.3.4 Powers in Three-Phase Systems

The process we used to calculate the power consumption in Examples 4.14.3 is rather cumbersome: we first had to figure out the voltage and current in each phase of the load, then calculate the power consumed in each phase and finally multiply by three. In practice this could even be impossible because we might not know whether a load is Y-connected, Δ-connected, or a parallel combination of the two. We need a uniform way to calculate the active and reactive powers using easily accessible quantities, i.e., the line-to-line voltage and the line current at the source.

The active and reactive powers consumed by a balanced three-phase load are given by:

$$P = 3\,V_{\text{phase}}\, I_{\text{phase}}\, \cos\theta$$
$$Q = 3\,V_{\text{phase}}\, I_{\text{phase}}\, \sin\theta \tag{4.22}$$

where V_{phase} and I_{phase} are, respectively, the voltage across and the current through the impedance in each phase and $\cos\theta$ is the power factor of the load. Let V_{LL} denote the line-to-line voltage and I_L the current in the line connecting the source to the load.

If this load is Y-connected, we have: $V_{\text{phase}} = \frac{1}{\sqrt{3}} V_{LL}$ and $I_{\text{phase}} = I_L$

On the other hand, if this load is Δ-connected, we have: $V_{\text{phase}} = V_{LL}$ and $I_{\text{phase}} = \frac{1}{\sqrt{3}} I_L$

Replacing V_{phase} and I_{phase} in terms of V_{LL} and I_L, we get expressions for the active and reactive powers that are valid for both Y-connected and Δ-connected loads:

$$P = \sqrt{3}\,V_{LL}\, I_L\, \cos\theta \tag{4.23}$$

$$Q = \sqrt{3}\,V_{LL}\, I_L\, \sin\theta \tag{4.24}$$

The apparent power in a three-phase system is then:

$$S = \sqrt{P^2 + Q^2} = \sqrt{3}\,V_{LL}\, I_L \tag{4.25}$$

Example 4.4 *Power in a three-phase load* The current drawn by a 10 kVA unity power factor three-phase load supplied from a 480 V source.

$$I_L = \frac{S}{\sqrt{3}\,V_{LL}} = \frac{10{,}000}{\sqrt{3}\times 480} = 12.03\,A$$

4.3.5 Single-Phase Representation and One-Line Diagrams

The previous sections make clear that, when we are dealing with a *balanced* three-phase system, we do not need to analyze separately what happens in each phase. The results for

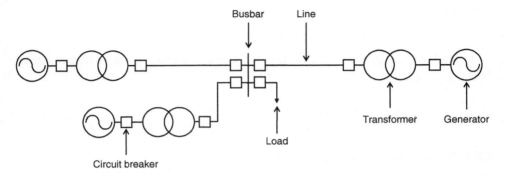

Figure 4.11 One-line diagram of a simple three-phase power system. Every item shown on this diagram represents a three-phase component.

one phase can be translated into the other phases through 120° shifts. We can take this further and treat balanced three-phase systems as if they were single-phase systems, as long as we use the formulas derived in Section 4.3.4 are used to calculate the powers.

One-line diagrams extend this concept to the graphical representation of power systems. In these schematic diagrams, such as the one shown in Figure 4.11, every item represents a three-phase component. There is no need for a "return conductor" because the currents in a balanced three-phase connection sum to zero.

4.4 Per Unit System

The analysis of three-phase power systems is further simplified if electrical quantities are normalized and expressed in per unit (pu). To normalize or calculate the per unit value of a quantity, we divide its magnitude in SI units by a base value for this quantity:

$$\text{per unit quantity} = \frac{\text{quantity in SI units}}{\text{base value of quantity in SI units}} \tag{4.26}$$

Per unit quantities are thus dimensionless and are denoted by the symbol pu. Only the magnitude of a complex quantity needs to be normalized because its angle is already dimensionless.

Example 4.5 *Per unit system* Suppose that the voltage at a particular node is $127 \angle 30°$ kV and that the base value for voltages at that node is 132 kV, the per unit value of this voltage is:

$$V^{\text{pu}} = \frac{127 \angle 30° \text{kV}}{132 \text{ kV}} = 0.962 \angle 30° \text{ pu}$$

We have to define base values for all the relevant electrical quantities: power, voltage, current, and impedance. While the choice of base values is in theory arbitrary, we must follow a few simple rules to maximize the benefits of the per unit system. We will first show how to choose the base values in single-phase power systems and then extend it to three-phase systems.

4.4.1 Choosing Base Quantities in Single-Phase Systems

The first step is to choose a common base value S_B for the active, reactive, complex, and apparent powers. This base power applies to the entire system under study.

We then choose the value of the nominal or rated voltage of each part of the system as the base voltage V_B for all components in this part of the system. If different parts of the system operate at different nominal voltages, they thus have different values of the base voltage. The advantage of this choice is that the per unit value of a voltage then clearly shows by how much the actual voltage deviates from the 1.0 pu nominal value.

We then choose the other base quantities in such a way that relations between electrical quantities have the same form whether they are expressed in per unit or in SI units.

In particular, in SI units the relationship between the apparent power and the magnitudes of the voltage and current is:

$$S = V I \tag{4.27}$$

In per unit, we want to be able to write:

$$S^{\text{pu}} = V^{\text{pu}} I^{\text{pu}} \tag{4.28}$$

which implies:

$$\frac{S}{S_B} = \frac{V}{V_B} \frac{I}{I_B} \tag{4.29}$$

Combining (4.27) and (4.29) gives the expression for the base current:

$$I_B = \frac{S_B}{V_B} \tag{4.30}$$

Similarly, Ohm's law in SI units is:

$$V = I Z \tag{4.31}$$

If we want to have the same expression in per unit:

$$V^{\text{pu}} = I^{\text{pu}} Z^{\text{pu}} \tag{4.32}$$

We must choose:

$$Z_B = \frac{V_B}{I_B} = \frac{V_B^2}{S_B} \tag{4.33}$$

Table 4.1 summarizes these choices and rules.

4.4.2 Choosing Base Quantities in Three-Phase Systems

The same principles apply to the choice of base quantities in three-phase systems, with a few twists to avoid having to make a distinction between line-to-line and line-to-neutral voltages and to get rid of pesky $\sqrt{3}$ factors. We again start by choosing a base power for the entire system and base voltages equal to the nominal voltages in each area of the system. To avoid having to specify whether a voltage is line-to-line or line-to-neutral, we would like their per unit values to be identical. In other words, we want:

$$V_{LL}^{\text{pu}} = V_{LN}^{\text{pu}} \tag{4.34}$$

Table 4.1 Rules governing the choice of base values in single-phase systems.

Quantity	Symbol	Rule
Apparent power	S_B	Arbitrary choice of one value for the entire system
Active power	P_B	$P_B = S_B$
Reactive power	Q_B	$Q_B = S_B$
Voltage	V_B	Nominal voltage at each location
Current	I_B	$I_B = \dfrac{S_B}{V_B}$
Impedance	Z_B	$Z_B = \dfrac{V_B^2}{S_B}$
Admittance	Y_B	$Y_B = \dfrac{1}{Z_B}$

Expressing (4.34) in terms of base quantities, we get:

$$\frac{V_{LL}}{V_{B,LL}} = \frac{V_{LN}}{V_{B,LN}} \tag{4.35}$$

Since in a balanced three-phase system we have:

$$V_{LN} = \frac{1}{\sqrt{3}} V_{LL} \tag{4.36}$$

We must therefore choose the base line-to-neutral voltage to be $\sqrt{3}$ times smaller than the base line-to-line voltage:

$$V_{B,LN} = \frac{V_{B,LL}}{\sqrt{3}} \tag{4.37}$$

For simplicity, we would like to be able to express the power in per unit as:

$$S^{\text{pu}} = V^{\text{pu}} I^{\text{pu}} \tag{4.38}$$

which implies:

$$\frac{S}{S_B} = \frac{V_{LL}}{V_{B,LL}} \frac{I_L}{I_B} \tag{4.39}$$

In SI units, the apparent power in a three-phase system is:

$$S = \sqrt{3} V_{LL} I_L \tag{4.40}$$

Inserting this expression in (4.39), we get:

$$\frac{\sqrt{3} V_{LL} I_L}{S_B} = \frac{V_{LL}}{V_{B,LL}} \frac{I_L}{I_B} \tag{4.41}$$

which determines the base current as a function of the base power and the base line-to-line voltage:

$$I_B = \frac{S_B}{\sqrt{3} V_{B,LL}} \tag{4.42}$$

In a balanced Y-connected load or in a balanced three-phase system where the loads have been combined into Y equivalents, we have:

$$V_{LN} = I_L Z_\phi \tag{4.43}$$

where Z_ϕ is the impedance in each phase. We would like the corresponding expression in per unit to be:

$$V_{LN}^{pu} = I_L^{pu} Z_\phi^{pu} \tag{4.44}$$

which implies that:

$$\frac{V_{LN}}{V_{B,LN}} = \frac{I_L}{I_B} \frac{Z_\phi}{Z_B} \tag{4.45}$$

Combining this expression with (4.43) and (4.42), we get:

$$Z_B = \frac{V_{B,LN}}{I_B} = \frac{V_{B,LL}^2}{S_B} \tag{4.46}$$

Table 4.2 summarizes the rules that must be followed when choosing base values in three-phase systems.

Example 4.6 *Base quantities* A small three-phase power system consists of two areas connected by a transformer. The nominal voltage is 132 kV in area Blue and 69 kV in area Red. Select base values for all quantities in both areas.

Let us choose $S_B = 100$ MVA. This choice applies to both areas.

The base line-to-line voltages are:

$$V_B^{Blue} = 132 \text{ kV}$$

$$V_B^{Red} = 69 \text{ kV}$$

Table 4.2 Rules governing the choice of base values in three-phase systems.

Quantity	Symbol	Rule
Apparent power	S_B	Arbitrary choice of one value for the entire system
Active power	P_B	$P_B = S_B$
Reactive power	Q_B	$Q_B = S_B$
Line-to-line voltage	$V_{B,LL}$	Nominal line-to-line voltage at each location
Line-to-neutral voltage	$V_{B,LN}$	$V_{B,LN} = \dfrac{V_{B,LL}}{\sqrt{3}}$
Current	I_B	$I_B = \dfrac{S_B}{\sqrt{3}\, V_{B,LL}}$
Impedance	Z_B	$Z_B = \dfrac{V_{B,LL}^2}{S_B}$
Admittance	Y_B	$Y_B = \dfrac{1}{Z_B}$

The base currents are:

$$I_B^{\text{Blue}} = \frac{S_B}{\sqrt{3}\, V_B^{\text{Blue}}} = 437.39\,A$$

$$I_B^{\text{Red}} = \frac{S_B}{\sqrt{3}\, V_B^{\text{Red}}} = 836.74\,A$$

The base impedances are:

$$Z_B^{\text{Blue}} = \frac{\left(V_B^{\text{Blue}}\right)^2}{S_B} = 174.24\,\Omega$$

$$Z_B^{\text{Red}} = \frac{\left(V_B^{\text{Red}}\right)^2}{S_B} = 47.61\,\Omega$$

Example 4.7 *Per unit currents* A load in area Blue of Example 4.6 draws 0.2 pu of apparent power at a voltage of 0.97 pu. Calculate the line current drawn by this load in Amps.

We are given: $S^{\text{pu}} = V^{\text{pu}}\, I^{\text{pu}} = 0.2$

Since $V^{\text{pu}} = 0.97$, we have:

$$I^{\text{pu}} = \frac{0.2}{0.97} = 0.206$$

Therefore:

$$I = I^{\text{pu}} \times I_B^{\text{Blue}} = 0.206 \times 437.39 = 90.184\,A$$

4.4.3 Converting Per Unit Impedances

Manufacturers specify the impedance of their equipment in per unit, where the base impedance is determined by the power and voltage rating of this equipment. If this equipment is to be considered in a system study that uses a different base power or base voltage, its per unit impedance must be converted to the new base. This conversion relies on the fact that the impedance value in ohms is invariant:

$$Z\,(\Omega) = Z_{\text{new}}^{\text{pu}} Z_{B,\text{new}} = Z_{\text{old}}^{\text{pu}} Z_{B,\text{old}} \tag{4.47}$$

Therefore:

$$Z_{\text{new}}^{\text{pu}} = Z_{\text{old}}^{\text{pu}} \frac{Z_{B,\text{old}}}{Z_{B,\text{new}}} \tag{4.48}$$

Expressing the old and new base impedances using (4.46), we have:

$$Z_{\text{new}}^{\text{pu}} = Z_{\text{old}}^{\text{pu}} \frac{\frac{V_{B,\text{old}}^2}{S_{B,\text{old}}}}{\frac{V_{B,\text{new}}^2}{S_{B,\text{new}}}} = Z_{\text{old}}^{\text{pu}} \frac{S_{B,\text{new}}}{S_{B,\text{old}}} \frac{V_{B,\text{old}}^2}{V_{B,\text{new}}^2} \tag{4.49}$$

Since it is unusual to connect equipment in an area of the system which does not have the same nominal voltage as their rating, this conversion usually simplifies to:

$$Z_{new}^{pu} = Z_{old}^{pu} \frac{S_{B,new}}{S_{B,old}} \tag{4.50}$$

Example 4.8 ***Base impedance conversion*** According to the data provided by its manufacturer, a device rated at 50 MVA has an impedance of 0.3 pu. This device must be studied as part of a system for which the base value for powers $S_B = 100$ MVA. What value of impedance should be used for these system studies?

According to (4.50), we have:

$$Z_{new}^{pu} = Z_{old}^{pu} \frac{S_{B,new}}{S_{B,old}} = 0.3 \times \frac{100}{50} = 0.6 \text{ pu}$$

In the remainder of this textbook, except where otherwise indicated, we will carry out calculations in per unit and will therefore omit the superscript pu.

4.5 Synchronous Generator Model

As we showed in Section 4.4.2, the rotation of the rotor magnetic field induces a set of three-phase voltages in the stator windings. Since these voltages are balanced, we can represent them by a single phasor $\overline{E}$, which we will call the induced or internal electromotive force (emf) of the generator. The magnitude of this emf is proportional to the magnitude of the rotor flux, which is a function of the dc current I_f flowing in the field winding:

$$|\overline{E}| = E = f(I_f) \tag{4.51}$$

If the stator windings are in open circuit, we can observe this voltage at the terminals of the generator. However, if the generator is connected to a load, the voltage $\overline{V}$ that we measure at these terminals will be affected by the impedance of the stator windings and the load current:

$$\overline{V} = \overline{E} - (R_S + jX_S)\overline{I} \tag{4.52}$$

where $\overline{I}$ is the phasor representing the stator currents, R_S is the resistance of the stator windings of the generator, and X_S their inductive reactance, which is called the synchronous reactance. Figure 4.12 illustrates this synchronous generator model. Since the magnitude

Figure 4.12 Single-phase steady state model of a synchronous generator.

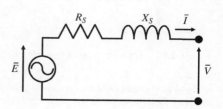

of R_S is considerably smaller than the magnitude of X_S, we will usually neglect it and use the following simplified expression in our analysis:

$$\overline{V} = \overline{E} - jX_S\overline{I} \tag{4.53}$$

4.6 Controlling an Isolated Synchronous Generator

The model developed in the previous section adequately describes the steady-state behavior of a generator from an electrical perspective. However, it is important to understand how the generator interacts with its mechanical prime mover and how it must be controlled to achieve desirable characteristics. To illustrate these concepts, consider a small portable generator like those that can be purchased from a hardware store. These devices are designed to provide ac power where or when it is not available from the grid. They combine a prime mover, typically an internal combustion engine, and a synchronous generator.

As the electrical load connected to its terminals varies, the generator should maintain a constant voltage magnitude and frequency at its terminals. Figure 4.13 illustrates the two control loops that are needed to achieve this goal. The first control loop measures the magnitude of the voltage V at the terminals of the generator and compares it to the reference value. If the measured voltage is less than the reference, the exciter increases the dc field current I_f, which increases the magnitude of the internal emf E, and restores the magnitude of V. Conversely, if the magnitude of V exceeds the reference, I_f and E are decreased.

The generator model of Figure 4.12 suggests that the power supplied to the load comes from the voltage source $\overline{E}$. However, we must keep in mind that this source represents the conversion of the mechanical power provided by the prime mover into electrical power. When the electrical load increases, the electrical power output from the generator exceeds the mechanical power provided by the prime mover. The difference is extracted from the kinetic energy of the mechanically coupled rotating masses of the prime mover and generator. As the prime mover and the rotor slow down,[1] the electrical frequency decreases.

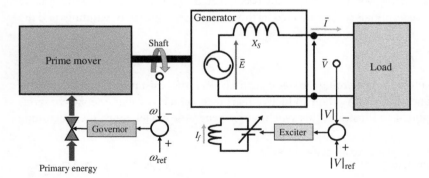

Figure 4.13 Schematic representation of the controls of an isolated synchronous generator.

---

1 If you have used one of these portable generators, you will have noticed how the noise it makes temporarily changes when you connect or disconnect a large load.

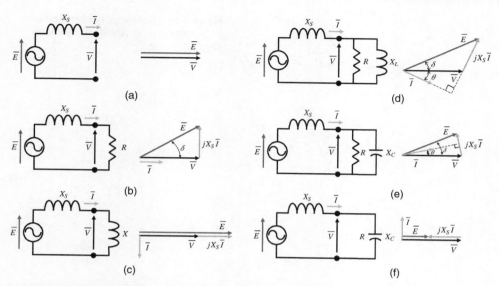

Figure 4.14 Operation of an isolated synchronous generator with different types of loads.

The second control loop of Figure 4.13 measures the rotational speed of the shaft connecting the prime mover and the generator and compares it to the rotational speed required to maintain nominal frequency. Acting on this difference, the governor increases the amount of primary energy (e.g., fuel) injected in the prime mover so that the amount of mechanical power it provides to the generator matches the electrical load. Conversely, if we decrease the electrical load on the generator, the rotational speed and frequency increases temporarily until the governor restores the power balance by reducing the fuel input.

These two control loops make it possible for such a generator to supply a variety of loads. Figure 4.14 shows the phasor diagram for each type of load. In each case, the phasor diagram results from the application of (4.53). The actions of the governor loop are implicit in these diagrams because it ensures that the frequency remains constant and that the necessary amount of active power is supplied. On the other hand, these diagrams show how the exciter loop adjusts the magnitude of the internal emf to maintain the terminal voltage V at a constant value. In particular, note the effect that an inductive and capacitive load have on the magnitude of $\overline{E}$. An inductive load requires a larger internal emf and thus a larger field current I_f, while this current must be reduced for a capacitive load.

4.7 Connecting a Generator to the Grid

Let us suppose that we want to connect a generator to a power systems that is already supplied by other generators. We first have to get the prime mover up and running. That process takes a few days for nuclear generators, a few hours for large steam plants, and a few minutes for open cycle gas turbines and diesel generators. Once the prime mover rotates the generator at close to its nominal speed, we can then turn on the excitation circuit and adjust the field current I_f until the voltage at the terminals of the generator is close to its nominal

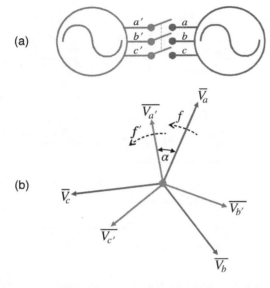

Figure 4.15 Generator synchronization: (a) Schematic representation; (b) Phasor diagram.

value. At that point we are in the situation illustrated in Figure 4.15a, where our generator is represented by the three-phase voltage source in blue, while the grid is modeled by the three-phase source in red. Figure 4.15b shows the phasors for the three-phase voltage produced by the generator in blue, and those of the grid in red. Several conditions must be satisfied before we can safely close the switch and connect the generator to the grid:

- The phase sequence must be the same on both sides. This means that the voltages reach their maximum value in the same order (a-b-c and a'-b'-c'). Swapping two of the leads (or in the lab reversing the direction of rotation of the generator) reverses the phase sequence.
- The rotational speed of the generator must be adjusted so that the frequency f' of the voltages produced by the generator matches the frequency f of the grid.
- The field current I_f must be adjusted so that the magnitude of the voltage at the terminals of the generator matches the grid voltage.
- The phase angle difference α between the generator and grid voltages must be minimized. This can be achieved by temporarily increasing or decreasing slightly the speed of the generator.

If these conditions are satisfied, closing the switch synchronizes the generator with the other generators connected to this system because the frequency at which it operates is locked to the grid frequency. Hence the name synchronous generator. On the other hand, if these conditions for synchronization are not satisfied, closing the switch creates a large and potentially damaging transient.

4.8 Operating a Synchronized Generator

We will discuss in subsequent chapters how the voltages and frequency in a power system are determined by the combined actions of all the generators and their interactions with the

Figure 4.16 Model of a generator connected to a grid represented as an ideal voltage source.

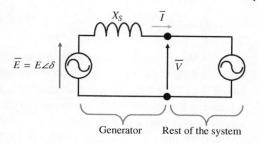

Generator Rest of the system

loads. In the meantime, let us consider a single generator synchronized to a large power system. We will assume that the power rating of this generator is a small fraction of the total power flowing through the system and is not able to affect the frequency and voltage of the grid. To reflect this assumption, we model the rest of the power system as an ideal voltage source connected directly to the terminals of the generator, as shown in Figure 4.16. Since we model the rest of the power system as an ideal voltage source, it maintains a constant voltage magnitude and frequency at the terminal of the generator. This model also emphasizes that instead of supplying a specific load, a synchronized generator injects active and reactive power into the grid.

Let us develop expressions showing how the active and reactive power produced by this generator are related to the magnitude and angle of the voltages of our generator model. The current flowing from the generator can then be expressed as:

$$\overline{I} = \frac{\overline{E} - \overline{V}}{jX_S} \tag{4.54}$$

The complex power produced by the generator is given by:

$$\overline{S} = \overline{V}\,\overline{I}^* = \overline{V}\frac{\overline{E}^* - \overline{V}^*}{-jX_S} \tag{4.55}$$

Using the voltage at the generator terminals as the reference for the angles and the following notations:

$$\overline{V} = V\angle 0° = V$$
$$\overline{E} = E\angle\delta = E\cos\delta + jE\sin\delta \tag{4.56}$$

Eq. (4.55) expands into:

$$\overline{S} = \frac{V(E\cos\delta - jE\sin\delta) - V^2}{-jX_S} \tag{4.57}$$

Multiplying the numerator and denominator by j and separating the real and imaginary parts, we get:

$$\overline{S} = P + jQ = \frac{EV\sin\delta}{X_S} + j\frac{VE\cos\delta - V^2}{X_S} \tag{4.58}$$

which gives us expressions for the active and reactive powers produced by the generator:

$$P = \frac{EV}{X_S}\sin\delta \tag{4.59}$$

$$Q = \frac{EV\cos\delta - V^2}{X_S}\tag{4.60}$$

Eq. (4.59) shows that, if we keep the magnitudes of $\overline{E}$ and $\overline{V}$ constant, the phase angle difference between these phasors increases with the amount of active power produced. δ is called the *power angle* of the generator. This angle should not be confused with the angle θ between the terminal voltage $\overline{V}$ and the current $\overline{I}$, which is the *power factor angle*. This equation also shows that there is an upper limit to the amount of power that a generator can deliver and that this limit occurs for $\delta = 90°$. However, under normal circumstances, the power angle δ is significantly smaller than $90°$. This relation is called the power angle curve and is illustrated in Figure 4.17.

If we neglect the losses in the generator, the electrical power that it produces is equal to the mechanical power provided by the prime mover. Given this mechanical power, we can calculate the angle δ_0 at which the generator is operating:

$$P_m = P = \frac{EV}{X_S}\sin\delta_0\tag{4.61}$$

Given δ_0 and the magnitude of the internal emf E, we can use (4.60) to calculate the amount of reactive power that the generator injects in the grid. If, as is normally the case, δ_0 is relatively small, and $\cos\delta_0$ is thus slightly less than 1.0, (4.60) then suggests that the reactive power injection is proportional to the difference in the magnitude of E and V. Adjusting the field current I_f therefore allows us to control this reactive power injection and make it positive or negative depending on the needs of the grid. Figure 4.18 illustrates how a generator can produce or absorb reactive power while producing the same amount of active power by keeping $E\sin\delta$ constant. When a generator produces reactive power, it is said to be overexcited. When it absorbs reactive power, it is described as underexcited. We will see in later chapters that reactive power injections play an important role in regulating voltages in the transmission network.

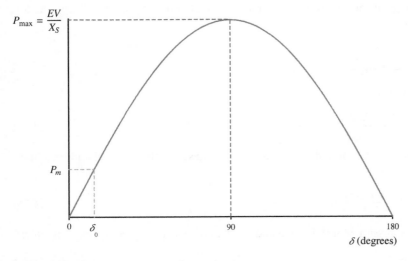

Figure 4.17 Power angle curve of a synchronous generator.

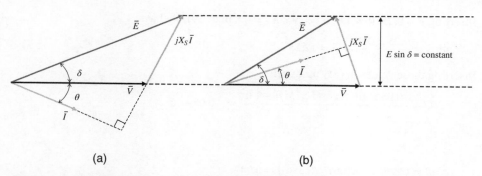

(a)   (b)

Figure 4.18 Operation of a synchronized generator when (a) producing reactive power and (b) absorbing reactive power.

Example 4.9 *Steady-state generator operation* The prime mover of a generator provides 0.25 pu of mechanical power. Its synchronous reactance has a value of 0.8 pu and the voltage V is at nominal value. The field current is such that the magnitude of the internal emf E is 1.15 pu. Calculate the power angle, the reactive power produced by this generator, the power factor, and the magnitude and angle of the current.

Inserting $P = 0.25$, $V = 1.0$, $E = 1.15$, $X_S = 0.8$ in (4.59), we have:

$$\sin \delta = \frac{0.25 \times 0.8}{1.0 \times 1.15} \rightarrow \delta = 10.01°$$

Inserting these values in (4.60), we get:

$$Q = \frac{[1.15 \times 1.0 \times \cos(10.01°)] - 1.0^2}{0.8} = 0.166 \text{ pu}$$

To calculate the power factor, we remember that:

$$\tan \theta = \frac{Q}{P}$$

Hence:

$$pf = \cos \theta = \cos[\tan^{-1}(0.662)] = 0.834$$

There are several ways of calculating the magnitude of the current. For example:

$$I = \frac{S}{V} = \frac{\sqrt{P^2 + Q^2}}{V} = \frac{0.3}{1.0} = 0.3 \text{ pu}$$

Combining with the result for the power factor and taking the terminal voltage $\overline{V}$ as the reference for the angles, we get:

$$\overline{I} = 0.3\angle - 33.5° \text{ pu}$$

The current lags the voltage because this generator supplies reactive power.

Example 4.10 *Effect of field current on generator operation* Repeat the calculation of the previous example for the case where the value of the field current is such that the magnitude of the internal emf is 0.95 pu.

Inserting $P = 0.25$, $V = 1.0$, $E = 0.95$, $X_S = 0.8$ in (4.59), we have:

$$\sin \delta = \frac{0.25 \times 0.8}{1.0 \times 0.95} \rightarrow \delta = 12.15°$$

Inserting these values in (4.60), we get:

$$Q = \frac{[0.95 \times 1.0 \times \cos(12.15°)] - 1.0^2}{0.8} = -0.089 \text{ pu}$$

$$\tan \theta = \frac{Q}{P}$$

$$pf = \cos \theta = \cos[\tan^{-1}(0.356)] = 0.942$$

$$I = \frac{S}{V} = \frac{\sqrt{P^2 + Q^2}}{V} = \frac{0.265}{1.0} = 0.265 \text{ pu}$$

Combining with the result for the power factor and taking the terminal voltage $\overline{V}$ as the reference for the angles, we get:

$$\overline{I} = 0.265 \angle 19.6° \text{pu}$$

The current leads the voltage because this generator absorbs reactive power.

4.9 Synchronous Condenser

The ability of a synchronous generator to control whether to produce or absorb reactive power is pushed to the extreme with synchronous condensers, which are synchronous generators that are not driven by a prime mover. Synchronous condensers are therefore not able to inject active power in the grid but can inject or absorb reactive power, as illustrated in Figure 4.19.

4.10 Generator Limits

We need to complement the generator model that we developed in Section 4.5 to reflect the fact that generators are not ideal voltage sources but physical devices whose operation is constrained.

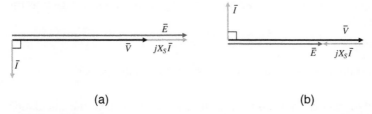

(a)   (b)

Figure 4.19 Operation of a synchronous condenser when (a) producing reactive power and (b) absorbing reactive power.

First, the amount of electric power that a generator can deliver in the steady state is limited by the mechanical power that its prime mover can provide:

$$P \leq P_{max} \tag{4.62}$$

The lower end of the range of mechanical power that a prime mover can provide varies depends on its nature. Diesel engines can provide just enough power to overcome mechanical losses and keep the generator rotating at synchronous speed. On the other hand, to ensure their safe and reliable operation, in the steady-state steam and gas turbines must provide a significant fraction of their maximum output. The extreme case being nuclear power plants, which are usually designed to operate at a constant power output. We can summarize these considerations as follows:

$$P \geq P_{min} \tag{4.63}$$

Even though it is small by design, the resistance of the stator windings creates losses proportional to the square of the current. The heat associated with these losses must be dissipated to prevent the temperature inside the stator to rise above a level where it would degrade the insulation and ultimately cause a fault. Because the amount of heat that can be dissipated is fixed, the stator current must be limited:

$$I \leq I_{max} \tag{4.64}$$

Since the operation of a generator is typically specified in terms of power rather than current, we need to express this constraint as a function of P and Q. The apparent power produced by a generator is related to the voltage and current by:

$$S = \sqrt{P^2 + Q^2} = VI \tag{4.65}$$

Combining (4.64) and (4.65), we get:

$$\frac{P^2 + Q^2}{V^2} \leq I_{max}^2 \tag{4.66}$$

For a constant value of the generator terminal voltage V, this constraint takes the form of a circle centered on the origin of the $P - Q$ plane, as shown in Figure 4.20.

In Section 4.8 we saw that the amount of reactive power produced by a generator increases with the rotor field current I_f. To avoid damages caused by excessive heat stemming for ohmic losses, this current must also be limited. This means that the internal emf of the generator cannot exceed a value E_{max}. To see how this constraint on the rotor field current limits the active and reactive power produced by the generator, we insert this value E_{max} in (4.59) and (4.60):

$$P = \frac{E_{max} V}{X_S} \sin \delta \tag{4.67}$$

$$Q + \frac{V^2}{X_S} = \frac{E_{max} V}{X_S} \cos \delta \tag{4.68}$$

Summing the square of these two equations, we get:

$$P^2 + \left(Q + \frac{V^2}{X_S}\right)^2 = \left(\frac{E_{max} V}{X_S}\right)^2 \tag{4.69}$$

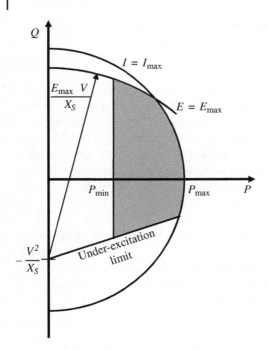

Figure 4.20 Loading capability diagram of a synchronous generator.

which is the equation of a circle or radius $\frac{E_{max}\, V}{X_S}$ centered at $\left(0, -\frac{V^2}{X_S}\right)$ in the $P - Q$ plane. Figure 4.20 shows the relevant portion of that circle.

It is generally undesirable for generators to absorbs large amounts of reactive power (negative Q) because this reduces the stability of the power system. Underexcitation must also be limited because it results in unacceptable heating. This under-excitation limit is sketched in Figure 4.20. The actual limit depends on the design of the generator and is provided by the manufacturer.

The shaded area in Figure 4.20 shows the range of active and reactive power that a generator can produce. Since the shape of this region is complex, the operating range of a generator is often expressed more simply as follows:

$$P_{min} \leq P \leq P_{max}$$
$$Q_{min} \leq Q \leq Q_{max} \tag{4.70}$$

Further Reading

Grainger, J. and Stevenson, W. (1994). *Power System Analysis*. McGraw Hill.
Glover, J.D., Overbye, T., and Sarma, M.S. (2016). *Power System Analysis and Design*, 6e. Cengage Learning.

Problems

P4.1 A synchronous generator has six poles. At what speed does it rotate when it generates a 60 Hz voltage? 50 Hz voltage?

P4.2 A three-phase voltage source is rated at 480 V. What is its nominal line-to-neutral voltage? What is its nominal line-to-line voltage?

P4.3 A balanced Y-connected three-phase load is supplied directly from a balanced 480 V three-phase source. Each phase of this load has an impedance $Z = 10 + j3\Omega$. Calculate the current in the connection between the source and the load, as well as the active and reactive powers supplied by this source to this load.

P4.4 A balanced Δ-connected three-phase load is supplied directly from a balanced 480 V three-phase source. Each phase of this load has an impedance $Z = 10 + j3\Omega$. Calculate the current in the connection between the source and the load, as well as the active and reactive powers supplied by this source to this load. Compare these results with those of problem P4.3.

P4.5 A balanced Y-connected three-phase load is supplied from a balanced 480 V three-phase source. Each phase of this load has an impedance $Z = 10 + j3\Omega$. The lines connecting this source to the load have an impedance $Z_l = 1 + j4\Omega$. Calculate the current in the lines connecting the source to the load and the active and reactive powers supplied by this source to this load.

P4.6 A balanced Δ-connected three-phase load is supplied from a balanced 480 V three-phase source. Each phase of this load has an impedance $Z = 10 + j3\Omega$. The lines connecting this source to the load have an impedance $Z_l = 1 + j4\Omega$. Calculate the current in the lines connecting the source to the load and the active and reactive powers supplied by this source to this load.

P4.7 A balanced 480 V three-phase source supplies a Y-connected load in parallel with a Δ-connected load. The impedance in each phase of the Y-connected load is $Z_Y = 20 + j7\Omega$ and the impedance in each phase of the Δ-connected load is $Z_\Delta = 24 + j9\Omega$. The lines connecting the source to these loads have an impedance $Z_l = 2 + j3\Omega$. Calculate the active and reactive powers supplied by this source as well as the losses in the lines.

P4.8 A three-phase load supplied from a 1 kV three-phase, 60 Hz source consumes 10 kW at 0.9 pf lagging. Model this load:
   (a) As a balanced Y-connected load where each phase consists of a resistance in series with an inductive reactance.

    (b) As a balanced Δ-connected load where each phase consists of a resistance in series with an inductive reactance.

    (c) As a balanced Y-connected load where each phase consists of a resistance in parallel with an inductive reactance.

    (d) As a balanced Δ-connected load where each phase consists of a resistance in parallel with an inductive reactance.

Calculate the values of these inductive reactances in Ω and in H.

P4.9 The nominal voltage in a given part of a three-phase power system is 132 kV. Given that the base apparent power is 100 MVA, determine:

    (a) The base line-to-line voltage.

    (b) The base line-to-neutral voltage.

    (c) The base current.

    (d) The base impedance.

P4.10 The manufacturer's data sheet for a 50 MVA 30 kV generator indicates that its synchronous reactance has a value of 0.8 pu. Calculate the per unit value of this synchronous reactance when the base apparent power is 100 MVA and

    (a) This generator is connected in a part of the network operating at its nominal voltage

    (b) This generator is connected in a part of the network operating at 25 kV.

P4.11 An isolated generator supplies at its nominal voltage a load that can be modeled as a resistance of 5 pu in parallel with an inductive reactance of 30 pu. Given that the synchronous reactance of this generator has a value of 0.8 pu, calculate the mechanical power provided by its prime mover and the magnitude and angle of its internal emf. Use the terminal voltage as the reference for the angles. Sketch a phasor diagram illustrating this operating condition.

P4.12 A generator operates at its nominal voltage and injects into the grid a current of 0.5 pu at 0.85 pf lagging. Given that its synchronous reactance is 0.7 pu, calculate the magnitude and phase of its internal emf and the active and reactive power produced by this generator. Sketch the phasor diagram for this operating condition. Use the terminal voltage as the reference for the angles.

P4.13 Repeat the previous problem for the case where the current is 0.5 pu at 0.9 pf leading.

P4.14 Given $S_B = 100$ MVA and $V_B = 30$ kV, express the answers of the previous two problems in SI units.

P4.15 A synchronous generator operates at its nominal voltage and injects into the grid 1.0 pu of active power and 0.3 pu of reactive power. Given that its synchronous reactance is 0.6 pu, calculate the magnitude and phase of its internal emf. Use the terminal voltage as the reference for the angles.

P4.16 A generator synchronized to the grid operates at its nominal voltage and supplies 1.0 pu of active power and −0.15 pu of reactive power. Given that its synchronous reactance is 0.6 pu, calculate the magnitude and phase of its internal emf. Use the terminal voltage as the reference for the angles.

P4.17 By how much should the internal emf of the generator of the previous problem be changed so that it operates at unity power factor?

P4.18 A generator synchronized to the grid operates at its nominal voltage. The field current is adjusted so that the magnitude of the internal emf is 1.4 pu. Given that the synchronous reactance has a value of 0.65 pu and that the generator operates at unity power factor, calculate the active power produced by this generator, the power angle, and the magnitude and phase of the stator current. Use the terminal voltage as the reference for the angles.

P4.19 A synchronous condenser has a synchronous reactance of 0.75 pu. Given that it operates at its nominal voltage, how much reactive power does it provide if:
(a) The magnitude of the internal emf is adjusted at 1.3 pu?
(b) The magnitude of the internal emf is adjusted at 0.8 pu?

P4.20 A balanced three-phase 25 kVA, 480 V, and 60 Hz generator is supplying a line current of 20 A per phase at a 0.8 lagging power factor at rated voltage. Determine the real and complex power supplied by the generator.

P4.21 A balanced Y-connected voltage source with $V_{an} = 277 \angle 0° \, V$ is connected directly to the parallel connection of a balanced Y-load and a balanced Δ-load, where $Z_Y = 20 + j10 \, \Omega$ and $Z_\Delta = 20 + j10 \, \Omega$. Using $S_{B, 1\emptyset} = 10 \, kVA$ and $V_{B, LN} = 277 \, V$, find the source current I_a in per-unit and in amperes. Find the total power consumed by the load. Hint: Convert the value of Z_Δ to a Y-connected impedance.

5

Electronic Power Conversion

5.1 Overview

In hydro and thermal power plants, the mechanical power delivered to the generator can be controlled quite precisely, making it possible to connect these generators directly to the rest of the power system. This is not the case with wind and photovoltaic generation because the amount of primary power available is variable and outside our control. Furthermore, PV panels convert solar energy directly into electrical energy, albeit as a dc voltage rather than an ac voltage. Power produced from these variable renewable energy sources must therefore be processed electronically before being injected into the power system. Power electronics is also required when energy must be extracted from one location, transmitted through a dc link, and injected at another location.

This chapter introduces the fundamental concepts of electronic power conversion as they apply to large power systems. We limit ourselves to a functional description of idealized power electronics converters. For a detailed discussion of the design and control of these converters, students should consult specialized power electronics textbooks, such as those listed under the heading Further Reading.

5.2 Switches

Energy efficiency requires that voltages and current be controlled in a discrete manner, i.e., either fully on or off. Power electronics therefore relies on devices that emulate on/off switches. These switching devices can be classified in terms of their directionality and controllability. A device is *bidirectional* if it allows the current to flow in either direction. It is *unidirectional* if it allows current to flow in only one direction. A *fully controllable* device can be opened and closed using a low-power control signal. An ideal switching device would be bidirectional and fully controllable. Since such a device has not yet been invented, the applications that we describe in this chapter have to be constructed using *reverse-conducting switching cells*. As Figure 5.1 shows, such cells consist of a fully controllable unidirectional switching device, for example an insulated-gate bipolar transistor (IGBT), in antiparallel with a diode. When a positive control signal is applied to the gate G, a current can flow

Power Systems: Fundamental Concepts and the Transition to Sustainability, First Edition. Daniel S. Kirschen.
© 2024 John Wiley & Sons Ltd. Published 2024 by John Wiley & Sons Ltd.
Companion website: www.wiley.com/go/kirschen/powersystems

through the IGBT from the collector C to the emitter E. This current is blocked when the control command is removed. If the device is reverse biased, a current can flow from the anode to the cathode of the diode, i.e., from A to K. We will neglect the voltage drops across these switching cells when they are conducting.

5.3 Voltage Source Converters

Controllable reverse-conducting switching cells make possible the development of voltage source converters that flexibly convert dc into ac or vice-versa. This section first describes the structure and operation of a single-phase converter and then briefly introduces three-phase converters.

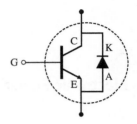

Figure 5.1
Reverse-conducting switching cell.

5.3.1 Single-Phase Voltage Source Converter

Figure 5.2 shows the circuit diagram of an H-bridge, which is a type of converter used to connect a dc subsystem to a single-phase ac subsystem. For now, we model the dc subsystem as an ideal dc voltage source and the ac subsystem as an ideal ac voltage source. An inductor connected in series with the ac subsystem ensures the continuity of the current i and prevents the switching cells from making a direct connection between incompatible voltages. It also allows the ac side of these converters to be current-controlled, while the dc side is voltage-controlled.

Let s_k denote the control signal applied to the gate of switching cell k. When $s_k = 1$, current can flow from through the transistor Q_k of switching cell k if the voltage between the collector to the emitter is positive. On the other hand, when $s_k = 0$ this current is blocked.

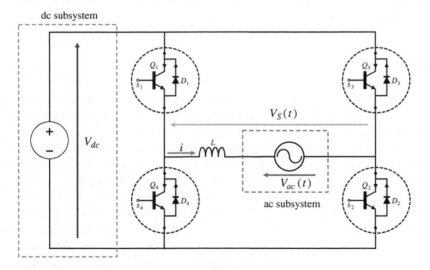

Figure 5.2 H-Bridge converter with ideal dc voltage source.

We clearly cannot have $s_1 = s_4 = 1$ or $s_3 = s_2 = 1$ because these combinations of control signals would cause a short circuit across V_{dc}. Instead, we control opposite pairs of switching cells in tandem by setting $s_1 = s_2 = 1$ while $s_3 = s_4 = 0$ or setting $s_1 = s_2 = 0$ while $s_3 = s_4 = 1$.

Let us examine what values the voltage V_S takes for these two combinations of control signals and the two possible directions of the current i.

1. $i > 0$; $s_1 = s_2 = 1$; $s_3 = s_4 = 0$
   i flows from the positive terminal of the dc source, through Q_1, L, the ac subsystem, and Q_2 to the negative terminal of the dc source. Since we neglect the voltage drops across the switching cells, we have $V_S = +V_{dc}$.
2. $i < 0$; $s_1 = s_2 = 1$; $s_3 = s_4 = 0$
   i flows from the negative terminal of the dc source, through D_2, the ac subsystem, L, and D_1 to the positive terminal of the dc source. Thus, we again have $V_S = +V_{dc}$.
3. $i > 0$; $s_1 = s_2 = 0$; $s_3 = s_4 = 1$
   i flows from the negative terminal of the dc source, through D_4, L, the ac subsystem, and D_3 to the positive terminal of the dc source. In this case, $V_S = -V_{dc}$.
4. $i < 0$; $s_1 = s_2 = 0$; $s_3 = s_4 = 1$
   i flows from the positive terminal of the dc source, through Q_3, the ac subsystem, L, and Q_4 to the negative terminal of the dc source, giving us again $V_S = -V_{dc}$.

The value of V_S is thus independent of the direction of the current i and uniquely determined by the control signals s_k. Figure 5.3 shows how we can create a square waveform $V_S(t)$ by alternating positive control signals on s_1, s_2 and s_3, s_4. Equalizing the durations of these two sets of positive signals produces an ac waveform with a zero average.

We control the frequency of $V_S(t)$ by adjusting the period T at which we alternate the control signals. We can also control the phase of $V_S(t)$ by adjusting the timing of these

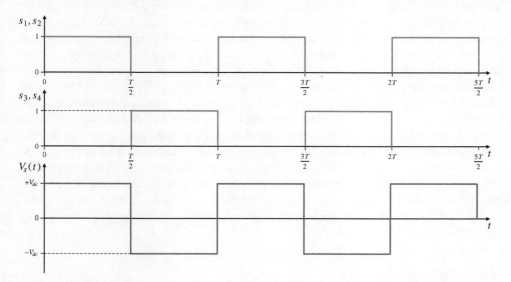

Figure 5.3 Relation between the control signals on the switching cells and the voltage $V_S(t)$.

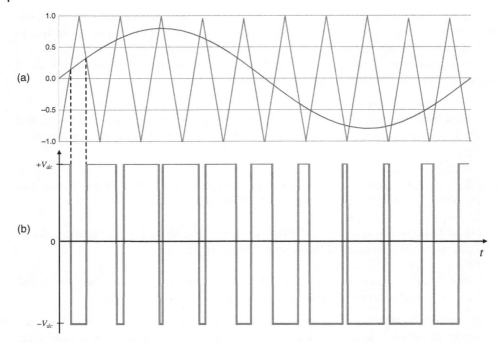

Figure 5.4 Pulse Width Modulation (PWM): (a) Triangular carrier and sinusoidal modulator; (b) Resulting PWM waveform.

control signals. We could change the magnitude of $V_S(t)$ by varying V_{dc}, but this is not always possible or practical. Instead, we control its magnitude and frequency using Pulse Width Modulation (PWM). Figure 5.4 illustrates a common form of PWM implemented by comparing two waveforms: a sinusoidal modulator and a higher frequency triangular carrier. The control signals of the switching cells are determined by the relative values of these two waveforms. When the value of the modulator is larger than the value of the carrier, the switching cells set $V_S(t) = +V_{dc}$. On the other hand, as illustrated by the vertical dashed lines, when the value of the carrier is larger than the value of the modulator, the switching cells set $V_S(t) = -V_{dc}$. The intersections between the carrier and the modulator thus create notches in $V_S(t)$. If we keep the amplitude of the triangular carrier constant, the width of these notches is controlled by the amplitude of the modulator. The resulting PWM waveform may not look like a sine wave, but a Fourier analysis shows that its fundamental frequency is equal to the frequency of the modulator and that the amplitude of its fundamental component increases with the amplitude of the modulator. In the remainder of our analysis of this type of converters, we will neglect the harmonics content of $V_S(t)$ and focus on its fundamental component $\widehat{V_S}(t)$.

5.3.2 Converter Operation

To make possible the conversion of power from dc to ac or vice versa, the frequency of $\widehat{V_S}(t)$ must be equal to frequency of ac subsystem. A phase-lock loop must therefore adjust the frequency of the PWM modulator to match the frequency of $V_{ac}(t)$. We therefore have:

$$\widehat{V_S}(t) = \widehat{V_S} \sin(\omega t + \delta) \tag{5.1}$$

and

$$V_{ac}(t) = V_{ac} \sin \omega t \tag{5.2}$$

Since both voltages have the same angular frequency ω, we can represent them by phasors:

$$\overline{V_S} = \frac{\widehat{V_S}}{\sqrt{2}} \angle \delta = E \angle \delta \tag{5.3}$$

$$\overline{V_{ac}} = \frac{V_{ac}}{\sqrt{2}} \angle 0° = V \angle 0° \tag{5.4}$$

where we have chosen the ac subsystem voltage as the reference for the angles and divided the magnitudes by $\sqrt{2}$ to follow the usual convention that phasors represent RMS quantities.

Referring to Figure 5.2, we observe that the voltages $V_S(t)$ and $V_{ac}(t)$ are separated by the inductor L. Figure 5.5 abstracts away the detail of the converter to focus on the relation between the phasor representation of these two voltages. We encountered a similar configuration in Figure 4.16 (Section 4.7) when we discussed the connection of a synchronous generator to the grid. We showed that the active power flowing across the inductor is given by:

$$P = \frac{EV}{X} \sin \delta \tag{5.5}$$

This expression is also applicable here and gives the amount of power that flows from the dc subsystem to the ac subsystem. It also suggests how we can regulate this flow. Assuming for now that the angular frequency ω and the voltage magnitudes E and V are constant, this power flow depends on the phase angle difference δ between these voltages. Since the control signals that determine $\widehat{V_S}(t)$ are synchronized with $V_{ac}(t)$ by the phase-lock loop, we can adjust their relative phase to achieve the desired power flow. If these signals are such that $\widehat{V_S}(t)$ leads $V_{ac}(t)$, δ is positive and the power flows from the dc subsystem to the ac subsystem. On the other hand, if we adjust these signals in such a way that $V_{ac}(t)$ leads $\widehat{V_S}(t)$, δ is negative and the power flows from the ac subsystem to the dc subsystem.

Figure 5.5 Connection between the ac voltage generated by the converter and the ac subsystem.

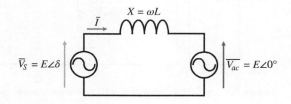

In Section 4.7, we also showed that in a system such as the one shown in Figure 5.5, the reactive power flowing from left to right is given by the following expression:

$$Q = \frac{EV \cos \delta - V^2}{X_S} \tag{5.6}$$

We can therefore regulate the reactive power produced or absorbed by this converter by controlling the magnitude E of the fundamental of the ac voltage generated by the converter.

We have assumed so far that the dc side could provide the amount of power controlled by the converter and injected on the ac side. This is often not the case. As we will see in Section 5.4, the amount of power available from the dc side is often variable. In such cases, we cannot model the dc subsystem as an ideal voltage source that maintains a constant voltage independently of the current or power that it supplies. Instead, as Figure 5.6 illustrates, we represent it by a source of dc power and add a capacitor designed to smooth the variations in the dc voltage. Neglecting losses in the converter, we can write the following power balance equation:

$$P_{dc}(t) = \frac{d}{dt} \left(\frac{CV_{dc}^2}{2} \right) + P_{ac}(t) \tag{5.7}$$

where $P_{dc}(t)$ is the variable power injected from the dc side, $P_{ac}(t)$ is the power converted into ac, and $\frac{CV_{dc}^2}{2}$ is the energy stored in the capacitor. $P_{ac}(t)$ must therefore be controlled to keep V_{dc} constant as $P_{dc}(t)$ fluctuates. This will ensure that the power output of the converter follows its power input.

5.3.3 Three-Phase Converter

For the sake of simplicity, we have based our explanations on the H-bridge converter which converts power between the dc side and a single-phase ac side. Figure 5.7 shows a converter configuration that interfaces a dc power source with a three-phase ac subsystem.

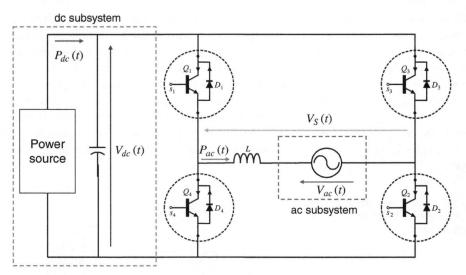

Figure 5.6 H-Bridge converter with dc power source.

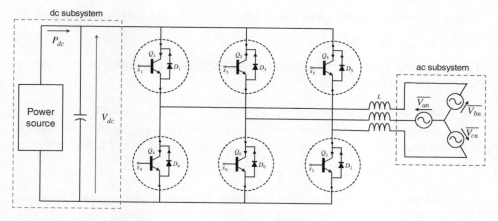

Figure 5.7 Three-phase dc/ac converter with dc power source.

5.4 Applications of Power Electronics in Power Systems

5.4.1 Battery Energy Storage

Charging and discharging batteries is conceptually the simplest application of power electronics converters in power systems. As a first approximation and as long as we respect their maximum and minimum state of charge and their current rating, batteries can be treated as ideal voltage sources. As Figure 5.8 illustrates, the converter can then be controlled to extract power from the grid to charge the battery or to extract power from the battery to inject in the grid. When discharging the battery, it operates as an inverter and produces ac from dc and. Neglecting losses, we have:

$$P_{dc} = P_{ac} = P_{ref} \geq 0 \tag{5.8}$$

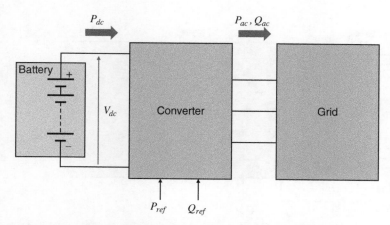

Figure 5.8 Schematic representation of the connection of a battery to the grid.

Conversely, when charging the battery, it acts as a rectifier and converts ac into dc:

$$P_{dc} = P_{ac} = P_{ref} \leq 0 \tag{5.9}$$

The rate at which the battery is charged or discharged must be limited to avoid prematurely degrading its energy capacity:

$$|P_{dc}| \leq P_{dc}^{max} \tag{5.10}$$

The converter can also produce or absorb reactive power. However, the active and reactive powers that a converter can produce or absorb are limited jointly by its current rating:

$$|S| = \sqrt{P_{ac}^2 + Q_{ac}^2} = V_{ac}I_{ac} \leq V_{ac}I_{ac}^{max} \tag{5.11}$$

Hence the maximum amount of active power that can be injected or extracted is reduced if the converter operates at a non-unity power factor:

$$P_{ac}^{max} = \sqrt{\left(V_{ac}I_{ac}^{max}\right)^2 - Q_{ac}^2} = V_{ac}I_{ac}^{max}\cos\theta \tag{5.12}$$

Example 5.1 *Effect of Power Factor* If a 5 kVA, 220 V converter operates at unity power factor, it can supply up to 5 kW of active power. If it operates at 0.9 pf, it can deliver up to 4.5 kW while absorbing or supplying 2.179 kvar.

5.4.2 PV Generation

Figure 5.9 illustrates the connection of a photovoltaic array to the grid. In this case, since the active power always flows from the dc side to the ac side, the converter is usually described as an inverter.

The PV characteristic of a solar array can be built based on the characteristics of individual cells such as those given in Figure 3.19. Figure 5.10 illustrates such a characteristic and shows that there is an optimal value V_{dc}^* that extracts the maximum amount of power from

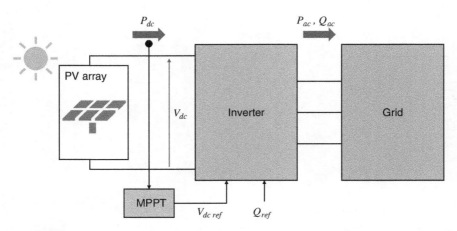

Figure 5.9 Schematic representation of the connection of a photovoltaic array to the grid.

this array. Unfortunately, as can be seen from Figure 3.19, this optimal voltage depends on the irradiance. It also depends on the temperature of the PV panels. Since both the irradiance and the temperature are unknown and vary continually, the voltage V_{dc} must track this optimal value if we want our system to maximize its power output under a wide range of operating conditions. The most common maximum power point tracking (MPPT) technique uses a "perturb-and-observe" algorithm that works as follows:

1. Let P_{dc}^{old} be the measured value of P_{dc} at the current V_{dc}
2. Using the control of the converter, set $V_{dc} = V_{dc} + \Delta V$
3. Let P_{dc}^{new} be the measured value of P_{dc} at this new V_{dc}
4. If $P_{dc}^{new} < P_{dc}^{old}$ set $\Delta V = -\Delta V$ (i.e., change direction)
5. Wait a short time
6. Go back to step 2

As Figure 5.10 illustrates, this algorithm adjusts V_{dc} to make the operating point "climb the hill." The operating point then oscillates around the peak as long as the irradiance and temperature do not change. When one of these does change, the algorithm detects that V_{dc} is no longer optimal and adjusts it up or down accordingly.

The active ac power P_{ac} injected by the inverter into the grid is thus driven by the power supplied by the PV array P_{dc}. Subject to the limit expressed by (5.11), the converter can also inject or absorb reactive power.

5.4.3   Type 4 Connection of a Wind Turbine

We saw in Chapter 3 that the rotational speed of a wind turbine must be adjusted as the wind speed changes to capture the maximum amount of wind power. In Chapter 4 we showed that the rotational speed of a synchronous generator connected to the grid is determined by the frequency of the grid and the number of poles of the generator. If we mechanically connect a wind turbine to a synchronous generator that is electrically connected directly to the grid, the rotational speed of the wind turbine would therefore be fixed, and we would not be able to adjust it as the wind speed changes. This section discusses how we can get around this problem using power electronics converters.

Figure 5.11 illustrates what is known as a Type 4 wind turbine connection. In this scheme, the wind turbine is connected to the rotor of the synchronous generator either directly or through a mechanical gearbox. The angular frequency ω_v of the voltage produced by

Figure 5.10   Maximum power point tracking using the perturb-and-observe technique.

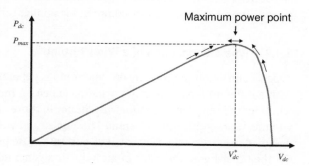

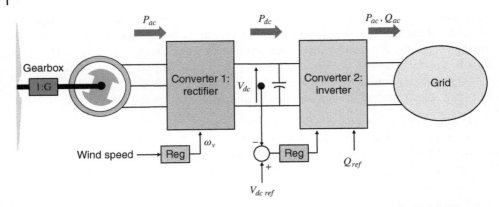

Figure 5.11 Schematic representation of the type 4 connection of a wind turbine to the grid.

this synchronous generator thus varies with the rotational speed of the turbine and is given by:

$$\omega_v = 2\pi \frac{p}{2} G \frac{N}{60}$$

(5.13)

where:

p is the number of poles of the synchronous generator
G is the gearbox ratio ($G = 1$ in a direct-drive turbine, i.e., without a gearbox)
N is the rotational speed of the turbine in rpm.

The first converter operates as a rectifier and converts the ac power produced by the generator into dc power. The second converter operates as an inverter and converts this dc power into ac power at the frequency of the grid.

The first converter controls the speed of the wind turbine to maximize the extraction of wind power. By controlling the voltage on the dc link between the two converters, the second converter ensures that the power injected in the grid is equal to the power extracted from the wind (assuming that the losses in the generator and the converters are negligible). The control signals of the inverter can also be adjusted to make it produce or absorb reactive power.

Type 4 connections provide excellent controllability and have become the preferred technology for the largest wind turbines. However, a Type 4 connection is expensive because it requires two converters designed to handle the full power rating of the turbine. For this reason, Type 3 connections are common for smaller turbines.

5.4.4 Type 3 Connection of a Wind Turbine

Type 3 connections do not rely on synchronous generators but instead generate electricity with a different type of rotating machine called an induction generator. While the stator windings of induction generators are similar to those of synchronous generators, the rotor windings are significantly different. The single rotor winding of a synchronous generator is designed to carry a relatively small dc current whose purpose is to create a magnetic field. On the other hand, the rotor of an induction generator holds three-phase windings that not only create a magnetic field but can also carry a substantial amount of power. Using slip rings that provide access to these rotating windings, we can inject or extract power to or from the rotor. Another important difference is that the rotor speed of an induction generator is not linked by a constant ratio to the frequency of the voltage applied to the stator. If we define the synchronous speed as follows:

$$\omega_s = 2\pi f \tag{5.14}$$

where f is the frequency of the voltage applied to the stator and ω_s is expressed in electrical angles, the mechanical speed of the rotor is:

$$\omega_m = (1 - s)\omega_s \tag{5.15}$$

where s is called the slip. If $s > 0$, the rotor rotates more slowly than synchronous speed and if $s < 0$, it rotates faster than synchronous speed. It can be shown that:

- The angular frequency of the currents flowing in the rotor windings is given by:

$$\omega_r = s\omega_s \tag{5.16}$$

- The amount of power flowing in or out of the rotor winding determines the slip, and hence the speed of the wind turbine. If we inject power into the rotor, the slip will be positive, and the turbine rotates at less than synchronous speed. Conversely, if we extract power from the rotor, the slip is negative, and the turbine rotates at faster than synchronous speed.

Figure 5.12 shows the components of a Type 3 connection. The stator of the induction generator is connected directly to the grid. Its synchronous speed is therefore determined by the operating frequency of the grid. A gearbox between the wind turbine and the induction generator raises the rotor speed to a value close to the synchronous speed. A cascade connection of two converters and a transformer links the rotor windings to the grid. The rotor-side converter converts variable frequency ac power from the rotor windings into dc. The grid-side converter converts this dc power into grid frequency ac power. The transformer raises the

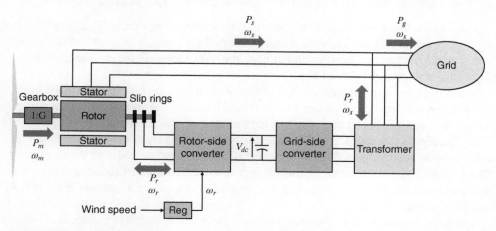

Figure 5.12 Schematic representation of the type 3 connection of a wind turbine to the grid with a doubly-fed induction generator.

output voltage of this converter to the grid voltage. This arrangement is called a doubly fed induction generator (DFIG) because both the stator and rotor are connected to the grid.

The wind turbine provides an amount P_m of mechanical power. Let P_s be the electrical power flowing out of the stator windings and P_r be the electrical power flowing in or out of the rotor windings. If we neglect the losses in the generator and the converters, the amount of power injected into the grid is:

$$P_m = P_g = P_s + P_r \tag{5.17}$$

where P_r is positive when the rotor-side converter extracts power from the rotor winding and negative when power is injected into the rotor winding. As mentioned above, controlling P_r allows us to control the mechanical speed of the rotor, and hence the speed of the wind turbine. This way, we can maximize the capture of wind power or keep the turbine speed constant as the wind power varies.

Since a large part of the wind power flows directly from the stator to the grid, Type 3 connections have the advantage over Type 4 connections that the converters must be designed to carry only a fraction of the rated power of the turbine. They are therefore quite common in medium size wind turbines.

Type 2 and Type 1 are simpler connection schemes used for smaller wind turbines. Since they are less efficient and controllable, they are rapidly becoming obsolete and will not be discussed here.

5.4.5 High-Voltage dc Links (HVDC)

Besides the wind turbine connections described in the previous sections, converting ac power into dc and then back to ac is necessary or desirable in the following situations:

- When power must be transferred between two power systems operating at different frequencies
- When power must flow through a long undersea transmission cable
- When a large amount of power must be transmitted over a long distance.

The purpose of these dc links is thus to extract power from one location in an ac grid and inject it at another. Power electronics converters make it possible to control directly and precisely the amount of power flowing through the link. As we will see in Chapter 8, this is a significant advantage over ac links.

In recent years, improvements in the voltage and current ratings of IGBTs have made possible the construction of voltage source HVDC links rated at 2000 MW. Higher ratings are possible using the older and less flexible technology of current source converters.

As Figure 5.13 illustrates, a HVDC system consists of two converters and a dc link. The two converters are connected to two different grids or to separate nodes in the same grid. The dc link can be either a long transmission line or cable, or a short connection between two converters located in the same building. In this latter case, such a system is described as a back-to-back dc system.

One of the converters, Converter 1 in the case of Figure 5.13, controls the amount of active power that it injects or extracts from the grid or node to which it is connected. The direction of the power flow determines the direction of the current in the dc link. With voltage source

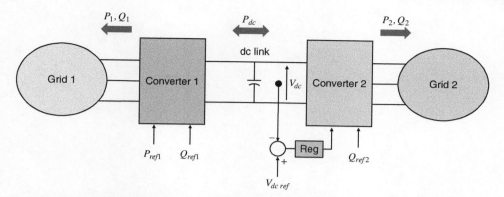

Figure 5.13 Schematic representation of a high voltage dc system built using voltage source converters.

technology, the other converter regulates the voltage across the dc link and thus ensures that this same amount of power, plus or minus losses, is extracted or injected from the other node or grid. Voltage source technology has the advantage that both converters can inject or extract reactive power from their respective grids.

Further Reading

Krein, P. (2014). *Elements of Power Electronics*, 2e. Oxford University Press.

Erickson, R.W. and Maksimovic, D. (2020). *Fundamentals of Power Electronics*, 3e. Springer.

Yazdani, A. and Iravani, R. (2010). *Voltage-Sourced Converters in Power Systems*. Wiley.

6

Balancing Load and Generation

6.1 Overview

As we discussed briefly in Chapter 1, power systems must maintain a balance between load and generation. In the first part of this chapter, we first address this issue from a physical perspective and show how it is related to the control of the frequency in the system. We then consider the question from a cost perspective and introduce the economic dispatch and unit commitment problems. Finally, we discuss how various sources of uncertainty affect this balance and the measures that must be taken to deal with these uncertainties.

6.2 Power Balance

Figure 6.1 shows a generic energy conversion system. The input to this system is primary energy from renewable or non-renewable sources. This energy is either converted to a different form of energy and output, stored internally, or dissipated in losses. The principle of conservation of energy requires that:

$$E_{in} = E_{out} + E_{losses} + E_{stored} \tag{6.1}$$

Taking the derivative of Eq. (6.1) with respect to time expresses this principle in terms of the input, output, lost and stored powers:

$$P_{in} = P_{out} + P_{losses} + P_{stored} \tag{6.2}$$

When analyzing a power system, we usually define the output as the set of all the nodes where loads are connected. P_{out} then represents the total electrical load. If the power input is exactly equal to the sum of the power output and the losses, $P_{stored} = 0$ and the energy stored in the system E_{stored} remains constant as long as this condition holds. Suppose that $P_{out} + P_{losses}$ increases suddenly and that P_{in} does not immediately follow this upsurge in load. To satisfy the power balance Eq. (6.2), we must have $P_{stored} < 0$, which means that the energy stored in the system decreases. Conversely, if $P_{out} + P_{losses}$ decreases suddenly or P_{in} increases while $P_{out} + P_{losses}$ remains constant, the energy stored in the system increases.

Power Systems: Fundamental Concepts and the Transition to Sustainability, First Edition. Daniel S. Kirschen.
© 2024 John Wiley & Sons Ltd. Published 2024 by John Wiley & Sons Ltd.
Companion website: www.wiley.com/go/kirschen/powersystems

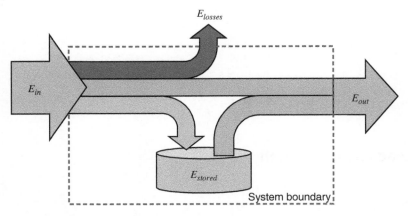

Figure 6.1 A generic energy conversion system.

6.3 Single Generator

Let us apply the concept of power balance to the system shown in Figure 6.2. where a single synchronous generator supplies an electrical load. If we draw the input boundary of the system across the shaft that connects the prime mover to the generator, P_{in} is then the mechanical power P_m transferred by this shaft. P_{out} is the electrical power P_e delivered by this generator to the load. Since this system involves a rotating mass, it stores energy in kinetic form:

$$E_{stored} = E_{kinetic} = \frac{1}{2} J \omega_m^2 \qquad (6.3)$$

where J is the combined moment of inertia of the generator and its prime mover, and ω_m is the mechanical rotational speed of this assemblage. P_{stored} is then:

$$P_{stored} = \frac{dE_{stored}}{dt} = J\omega_m \frac{d\omega_m}{dt} \qquad (6.4)$$

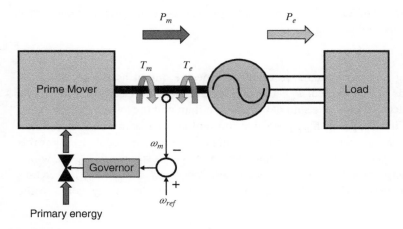

Figure 6.2 Isolated generator supplying a load.

Any imbalance between P_m and P_e will therefore cause an acceleration or a deceleration of the rotational speed ω_m. To get a deeper insight in the dynamics of this rotational system, we write its Newton's law of motion:

$$J\frac{d^2\theta_m}{dt^2} = T_m - T_e \tag{6.5}$$

where:

θ_m is the angular position of the rotor. The time derivative of this position is the speed of the rotor:

$$\frac{d\theta_m}{dt} = \omega_m \tag{6.6}$$

T_m is the torque that the prime mover exerts on the shaft connecting it to the generator. Multiplying this torque by ω_m gives the mechanical power input:

$$P_m = \omega_m T_m \tag{6.7}$$

T_e is the reaction torque that the generator exerts on the shaft in the opposite direction. If we neglect the losses, and multiply this reaction torque by ω_m, we get the electrical power supplied to the load:

$$P_e = \omega_m T_e \tag{6.8}$$

Multiplying both sides of Eq. (6.5) by ω_m and using Eqs. (6.6)–(6.8) leads to:

$$J\omega_m\frac{d\omega_m}{dt} = P_m - P_e \tag{6.9}$$

When the mechanical power input P_m is equal to the electrical power output P_e, we have:

$$\frac{d\omega_m}{dt} = 0 \tag{6.10}$$

The speed of the generator and hence the frequency of the voltage supplied to the load are then constant. Our system is in the steady state.

If the electrical load suddenly increases from P_e to P'_e, we have:

$$J\omega_m\frac{d\omega_m}{dt} = P_m - P'_e < 0 \tag{6.11}$$

which means that the mechanical speed and the frequency decrease because the extra energy consumed by the load is being extracted from the kinetic energy stored in the rotating mass. Obviously, this can't go on for too long; otherwise, the frequency would drop to an unacceptably low value. As Figure 6.2 illustrates, the prime mover is therefore equipped with a governor that takes as its input the difference between the reference speed ω_{ref}, which is the speed that corresponds to nominal frequency, and the actual speed ω_m. When this difference is positive, the governor injects more primary energy in the prime mover to increase its mechanical power output and restore the equilibrium with $P'_m = P'_e$. At that point the speed, frequency, and stored kinetic energy are again constant. Conversely, when the difference between the actual and reference speed is negative, the governor reduces the primary energy intake and hence the mechanical power input. The governor implements what is called *primary frequency control*.

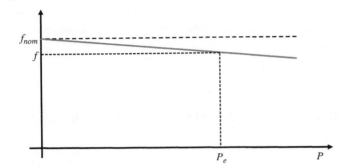

Figure 6.3 Generator droop characteristic.

For a reason that will become clear when we discuss what happens when the load is being supplied by multiple generators operating in parallel, the governor does not maintain the frequency at exactly its nominal value as the load varies. Instead, it implements a linear droop characteristic where the frequency must decrease slightly for the power output to increase. As Figure 6.3. illustrates, the operating frequency is determined by the generator droop characteristic and the electrical load that the generator supplies.

The droop coefficient ρ specifies by how much the frequency decreases from its nominal value when the power produced by the generator increases from zero to its maximum value. The mathematical formulation of the droop characteristic is thus:

$$\frac{f}{f^{nom}} = 1 - \rho\frac{P}{P^{max}} \tag{6.12}$$

or:

$$P = P^{max}\left(\frac{f^{nom} - f}{\rho f^{nom}}\right) \tag{6.13}$$

The droop coefficients ρ of generators typically range from 4% to 9%.

Example 6.1 *Droop coefficient* The governor of a 50 kW, 60 Hz generator has a droop coefficient of 4%. Calculate the frequency provided by this generator when it supplies a 25 kW load.

Inserting these values in Eq. (6.12), we get:

$$f = 60\left(1 - 0.04\frac{25}{50}\right) = 58.8\ \text{Hz}$$

6.4 Multiple Generators

Let us turn our attention to the case where the loads are supplied by multiple generators. If there are no battery energy storage devices, the total energy stored in the system is the sum of the kinetic energy of the rotating masses of all the generators and their prime movers:

$$E_{stored} = E_{kinetic} = \frac{1}{2}\sum_{g\in G} J_g\,\omega^2_{mg} \tag{6.14}$$

where G is the set of synchronous generators, J_g is the moment of inertia of generator g, and ω^2_{mg} is the square of the mechanical rotational speed of generator g.

In the steady state, the sum of the electrical powers produced by these generators is equal to the sum of the electrical powers consumed by all the loads and the losses in the network connecting these generators to these loads. In the steady state, if we neglect the losses in the generators, it is also equal to the sum of the mechanical powers supplied by the prime movers:

$$\sum_{g\in G} P_{m,g} = \sum_{g\in G} P_{e,g} = \sum_{l\in L} P_l + P_{\text{losses}} = P_T \tag{6.15}$$

where $P_{m,g}$ is the mechanical power provided by the prime mover of generator g; $P_{e,g}$ is the electrical power produced by the generator g; L is the set of loads; P_l is the power consumed by load l; P_{losses} is the losses in the network.

If the load increases suddenly, this equilibrium no longer holds. To restore the power balance the deficit must be extracted from the energy stored in the rotating masses. The rotational speed of the generators and hence the frequency of the system decrease according to the following equation:

$$J_G \omega_m \frac{d\omega_m}{dt} = \sum_{g\in G} P_{m,g} - \sum_{l\in L} P_l + P_{\text{losses}} < 0 \tag{6.16}$$

where J_G represents the combined inertia of all the generators and their prime movers. Conversely, if the load decreases, the system frequency will increase. While changes in load are typically small or slow, a much bigger disturbance to the equilibrium of Eq. (6.15) occurs when a large generator is suddenly disconnected from the system due to a fault, leading to a much more rapid decrease in frequency. Since the combined inertia of a system increases with the number and size of the generators, frequency drops due to a sudden loss of generation are much smaller in large interconnected systems than in small isolated systems.

Since all generators in a system are equipped with a governor that implements a droop characteristic like the one shown in Figure 6.3, they will all react to a drop in frequency by increasing their active power generation. An increase in load is thus shared by all the generators. Figure 6.4 illustrates this sharing in a system supplied by two generators.

Assuming they have the same droop coefficient, the powers produced by the two generators are given by:

$$P_1 = P_1^{\text{max}} \left(\frac{f^{\text{nom}} - f}{\rho f^{\text{nom}}} \right) \tag{6.17}$$

$$P_2 = P_2^{\text{max}} \left(\frac{f^{\text{nom}} - f}{\rho f^{\text{nom}}} \right) \tag{6.18}$$

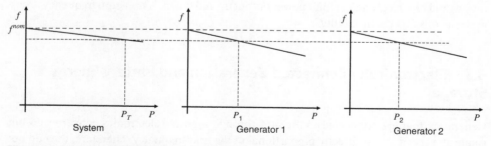

Figure 6.4 Load sharing using droop control in a two-generator system.

Since these two generators must supply the total system load P_T, we have:

$$P_T = P_1 + P_2 = \left(P_1^{max} + P_2^{max}\right) \left(\frac{f^{nom} - f}{\rho f^{nom}}\right) \tag{6.19}$$

From Eq. (6.19) we can determine the system operating frequency:

$$f = f^{nom} \left[1 - \rho \frac{P_T}{\left(P_1^{max} + P_2^{max}\right)}\right] \tag{6.20}$$

Inserting this value of the frequency in Eqs. (6.17) and (6.18) we get the power produced by each generator:

$$P_1 = \frac{P_1^{max}}{\left(P_1^{max} + P_2^{max}\right)} P_T \tag{6.21}$$

$$P_2 = \frac{P_2^{max}}{\left(P_1^{max} + P_2^{max}\right)} P_T \tag{6.22}$$

Example 6.2 *Parallel connection of two generators* Two 60 Hz generators supply a load of 20 kW. The first generator is rated at 20 kW and the second at 30 kW. They have the same droop coefficient of 0.04. Calculate the power produced by each generator and the operating frequency.

Using Eqs. (6.21) and (6.22), we get:

$$P_1 = \frac{20}{20 + 30} \times 20 = 8 \, kW$$

$$P_2 = \frac{30}{20 + 30} \times 20 = 12 \, kW$$

Equation (6.20) then gives:

$$f = 60 \left(1 - 0.04 \times \frac{20}{20 + 30}\right) = 59.04 \, Hz$$

The droop characteristic of the primary frequency control performed by the governors thus ensures that the load is shared by the generators. If they have the same droop coefficient, generators share the load in proportion to their maximum capacity. If this droop coefficient were set to zero, the generators would try to keep the frequency at its nominal value. However, because of inevitable small differences in the frequency they measure, they would repeatedly adjust their power output up and down. A non-zero frequency droop prevents this form of instability.

6.5 Electronically Connected Generation and Battery Energy Storage

Generators that are connected to the grid through a power electronics converter do not rotate at a speed that is directly proportional to the grid frequency. Therefore, they do not inherently contribute to maintaining the power balance by converting stored kinetic energy to and from electrical energy.

The controls of battery energy storage systems typically include a feature that detects rapid changes in grid frequency and triggers a temporary increase or decrease in power output. Storage devices make an increasingly important contribution to primary frequency control.

6.6 Secondary Frequency Control

The basic droop-based primary frequency control that we discussed in the previous sections provides an effective way of allocating changes in load among geographically distributed generators without requiring explicit communication between these generators. However, this primary frequency control does not:

- Maintain the frequency at or close to its nominal value.
- Ensure that the load is supplied in the most economical manner.

To remedy these shortcomings, most large power systems implement a centralized secondary frequency control mechanism on top of the primary frequency control. Instead of implementing a droop characteristic like the one shown in Figure 6.3, one can modify the governor to realize characteristics like the ones shown in Figure 6.5. Under this form of droop control, the power output P_i of generator i varies around a reference value P_i^{ref} based on a local measurement of the frequency f. To restore the frequency to its nominal value f^{nom}, the centralized secondary frequency control system periodically adjusts the reference values P_i^{ref} in such a way that their sum tracks the total load on the system:

$$\sum_{i \in G} P_i^{\text{ref}} \sim P_T \tag{6.23}$$

Note that in real life this total load is constantly changing and thus never known precisely. That is why the power balance is maintained through control of the frequency.

Since the generators usually have more than enough capacity to supply the load, there is more than one way to satisfy Eq. (6.23). This flexibility allows the secondary frequency control to choose the reference values P_i^{ref} that minimize the cost of producing the power needed to meet the load. The process of determining the optimal amount of power that each generator should produce is called economic dispatch and is discussed in the next section.

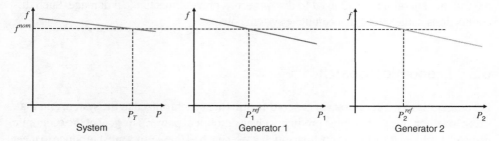

Figure 6.5 Secondary frequency control in a two-generator system.

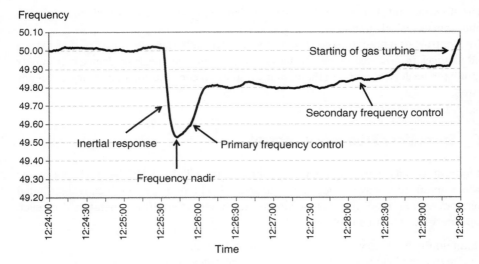

Frequency

Figure 6.6 Evolution of the system frequency during a major disturbances.

6.7 System Response to Large Disturbances

While the system inertia as well as the primary and secondary frequency control systems operate continuously in response to small fluctuations in the load, they must also be ready to counteract large disturbances, such as the sudden disconnection of a large generating unit. Figure 6.6 shows how the frequency evolved in a medium-size isolated system before and after the loss of a major generating unit. Prior to the event at 12 : 25 : 30, the frequency exhibited small random fluctuations around its nominal 50 Hz value. Immediately after the disconnection of this generating unit, the system frequency dropped very rapidly because of the relatively small total inertia of this system. Primary frequency control arrested this frequency decline at a nadir slightly above 49.50 Hz and brought it back to 49.80 Hz. Secondary response then further stabilized the frequency. At 12 : 29 : 30, the operator returned the frequency to its nominal value by starting up some gas turbines.

If the system frequency had dropped much below 49.50 Hz, underfrequency relays would have triggered the disconnection of loads in a last-ditch attempt at restoring the power balance and preventing a system collapse. If the frequency and the rotational speed decrease too much, generators are allowed to disconnect to prevent mechanical damage. Such disconnections usually lead to a complete system collapse.

6.8 Economic Dispatch

Let us assume that a certain set of generating units are synchronized to the system and that, taken together, these units must produce a given amount of power. The goal of the economic dispatch problem is to determine how much power each of these units must produce to meet this load at minimum cost. Specifically, we want to minimize the cost of the fuel needed to produce electrical power. For generation from renewable energy sources, this "fuel cost" is obviously zero. The output of these generators will therefore be taken as constant and at the maximum value determined by the availability of their source.

6.8.1 Heat Rate Curve and Cost Curve

The heat rate curve of a generating unit indicates how much primary energy must be injected in the form of fuel each hour to maintain a given electrical power output. Figure 6.7 shows the typical shape of such a curve. Most large thermal generating units cannot operate in the steady state below their minimum stable generation P^{min}. The heat rate tends to rise sharply as the power output approaches its maximum output P^{max}. The fuel input is measured in terms of its heat content and typically expressed in the practical unit of million British thermal units per hour (MBTU/h).

If we multiply the hourly fuel input by the cost of its fuel (in \$/MBTU), we get the curve of Figure 6.8 that gives the cost of operating this unit for an hour as a function of its electrical power output. These two curves have the same shape, but the heat rate curve changes only slightly as the power plant ages while the fuel cost depends on the often-volatile market price. Given the heat rate data and the fuel cost, the operating cost of each generating unit can be expressed as a convex quadratic or piecewise linear function of its power output. Note that an extrapolation of the cost curve to $P = 0$ usually leads to a non-zero value that is call the no-load cost of the generating unit.

Figure 6.7 Typical shape of the heat rate curve of a thermal generating unit.

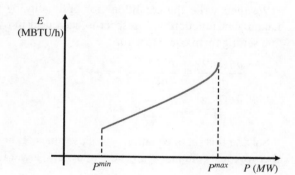

Figure 6.8 Typical shape of the cost curve of a thermal generating unit.

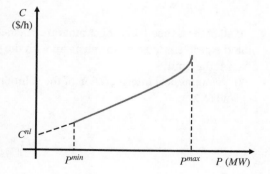

6.8.2 Mathematical Formulation

To formulate the economic dispatch as a mathematical optimization problem, we must define the decision variables, the objective function, and the equality and inequality constraints. In this case, the decision variables are the active power output P_i of each generating unit of the controllable set G. The objective is to minimize the total cost of operating the system. Given the cost function $C_i(P_i)$ of each generating unit, the objective function can be written as follows:

$$C^{\text{total}} = \sum_{i \in G} C_i(P_i) \tag{6.24}$$

To ensure that the system remains in equilibrium as we adjust the output of the generating units, we add an equality constraint to enforce the balance between the load L and the total generation:

$$\sum_{i \in G} P_i = L \tag{6.25}$$

To solve such a constrained optimization problem, we first construct its Lagrangian function $\mathcal{L}$ by adding to the objective function the equality constraint multiplied by an unknown variable λ called the Lagrange multiplier:

$$\mathcal{L}(P_i, \lambda) = \sum_{i \in G} C_i(P_i) + \lambda \left(L - \sum_{i \in G} P_i \right) \tag{6.26}$$

We then write the conditions for optimality by taking the partial derivatives of the Lagrangian function with respect to each decision variable and the Lagrange multiplier and setting them equal to zero:

$$\frac{\partial \mathcal{L}(P_i, \lambda)}{\partial P_i} \equiv \frac{dC_i(P_i)}{dP_i} - \lambda = 0 \; \forall i \in G \tag{6.27}$$

$$\frac{\partial \mathcal{L}(P_i, \lambda)}{\partial \lambda} \equiv L - \sum_{i \in G} P_i = 0 \tag{6.28}$$

Solving this set of equations gives us values of the power output of the generating units that we must compare against the upper and lower limits of each unit:

$$P_i \geq P_i^{\min} \; \forall i \in G \tag{6.29}$$

$$P_i \leq P_i^{\max} \; \forall i \in G \tag{6.30}$$

If all of these inequality constraints are satisfied, we have found the economic dispatch. If not, the problem must be solved again with the power outputs that are outside the bounds fixed at the limit.

To get a physical interpretation of the solution of the economic dispatch problem, let us expand (6.27):

$$\frac{dC_1(P_1)}{dP_1} = \frac{dC_2(P_2)}{dP_2} = \cdots = \lambda \tag{6.31}$$

$\dfrac{dC_i(P_i)}{dP_i}$ is the derivative of the cost of production of unit i. It represents the rate at which this cost increases with an increase in the output of this unit. If we consider a sufficiently small increase ΔP_i in the power output, we can make the following approximation:

$$\frac{dC_i(P_i)}{dP_i} \approx \frac{\Delta C_i}{\Delta P_i} \qquad (6.32)$$

or:

$$\Delta C_i = \frac{dC_i(P_i)}{dP_i} \Delta P_i \qquad (6.33)$$

$\dfrac{dC_i(P_i)}{dP_i}$ can thus be interpreted as the cost of producing one more MW from unit i, or the savings that would result from producing one less MW from this unit. It is called the *incremental or marginal cost* of unit i. Eq. (6.27) or (6.31) thus expresses the fact that at the optimum all the generating units should operate at the same marginal cost, i.e., that the cost of producing an extra MW from any of these units must be the same. Eq. (6.31) also provides a physical interpretation for the Lagrange multiplier λ: it represents the cost of supplying one more MW of load and is referred to as the *system marginal cost*. Since typically the cost functions increase monotonically, the system marginal cost increases with the load.

Example 6.3 *Economic dispatch* Three generating units supply the load of a small power system. Their cost functions (in \$/MWh) are:

$$C_1(P_1) = 2100 + 8P_1 + 0.1\,P_1^2$$

$$C_2(P_2) = 7200 + 7P_2 + 0.06\,P_2^2$$

$$C_3(P_3) = 6250 + 9P_3 + 0.07\,P_3^2$$

Neglecting for now the upper and lower limits on the output of these units, let us calculate the optimal economic dispatch of these generating units when they supply a load $L = 800$ MW and a load $L = 1200$ MW.

The Lagrangian of this optimization problem is:

$$\mathcal{L}(P_1, P_2, P_3, \lambda) = 2100 + 8P_1 + 0.1\,P_1^2 + 7200 + 7P_2 + 0.06\,P_2^2 + 6250 + 9P_3$$
$$+\, 0.07\,P_3^2 + \lambda(L - P_1 - P_2 - P_3)$$

Taking the partial derivatives of this Lagrangian gives the necessary optimality conditions:

$$\frac{\partial \mathcal{L}(P_1, P_2, P_3, \lambda)}{\partial P_1} \equiv \frac{dC_1(P_1)}{dP_1} - \lambda = 8 + 0.2P_1 - \lambda = 0$$

$$\frac{\partial \mathcal{L}(P_1, P_2, P_3, \lambda)}{\partial P_2} \equiv \frac{dC_2(P_2)}{dP_2} - \lambda = 7 + 0.12P_2 - \lambda = 0$$

$$\frac{\partial \mathcal{L}(P_1, P_2, P_3, \lambda)}{\partial P_3} \equiv \frac{dC_3(P_3)}{dP_3} - \lambda = 9 + 0.14P_3 - \lambda = 0$$

$$\frac{\partial \mathcal{L}(P_1, P_2, P_3, \lambda)}{\partial \lambda} \equiv L - P_1 - P_2 - P_3 = 0$$

Using the first three optimality conditions to express P_1, P_2, and P_3 as a function of λ, we get:

$$P_1 = 5.00\lambda - 40.00$$

$$P_2 = 8.33\lambda - 58.33$$

$$P_3 = 7.14\lambda - 64.29$$

Inserting these values in the fourth optimality condition gives us the optimal value of λ:

$$\lambda = \frac{L + 162.619}{20.476}$$

For $L = 800$ MW, the solution of the economic dispatch is:

$$\lambda = 47.012 \, \$/\text{MWh}$$

$$P_1 = 195.06 \, \text{MW}; P_2 = 333.43 \, \text{MW}; P_3 = 271.51 \, \text{MW}$$

All three generating units operate at the same marginal cost, which tells us how much one extra MW of load would cost each hour:

$$\frac{dC_1(P_1)}{dP_1} = \frac{dC_2(P_2)}{dP_2} = \frac{dC_3(P_3)}{dP_3} = \lambda = 47.012 \, \$/\text{MWh}$$

We can also calculate the total cost of generation at each hour:

$$C = C_1(P_1) + C_2(P_2) + C_3(P_3) = 37{,}523.69 \, \$/\text{h}$$

For $L = 1200$ MW, the solution of the economic dispatch is:

$$\lambda = 66.547 \, \$/\text{MWh}$$

$$P_1 = 292.73 \, \text{MW}; P_2 = 496.22 \, \text{MW}; P_3 = 411.05 \, \text{MW}$$

In this case, one extra MW of load would cost at each hour:

$$\frac{dC_1(P_1)}{dP_1} = \frac{dC_2(P_2)}{dP_2} = \frac{dC_3(P_3)}{dP_3} = \lambda = 66.547 \, \$/\text{MWh}$$

Each hour, producing this amount of power costs:

$$C = C_1(P_1) + C_2(P_2) + C_3(P_3) = 60{,}235.32 \, \$/\text{h}$$

Example 6.4 *Economic dispatch and renewable generation* Suppose that in addition to the three generating units of Example 6.3, this system is also supplied from a renewable energy source, which is expected to produce 200 MW. If we assume that the operating cost of this renewable source is negligible, we can simply subtract its production from the load, which gives us the net load that must be produced by the dispatchable thermal generating units. For $L = 800$ MW, $L_{net} = 800 - 200 = 600$ MW.

Using the intermediate results of the previous example, we get:

$$\lambda = 37.244 \text{ \$/MWh}$$

$$P_1 = 146.22 \text{ MW}; P_2 = 252.03 \text{ MW}; P_3 = 201.74 \text{ MW}$$

$$C = C_1(P_1) + C_2(P_2) + C_3(P_3) = 29{,}098.11 \text{ \$/h}$$

Example 6.5 *Effect of generator limits on economic dispatch* In the previous examples, we ignored the fact that there are upper and lower limits on the power that a generating unit can produce. Let us revisit Example 6.3 considering these limits:

$$200 \leq P_1 \leq 500$$

$$200 \leq P_2 \leq 450$$

$$150 \leq P_3 \leq 500$$

For $L = 1200$ MW, we found the following unconstrained economic dispatch solution:

$$P_1 = 292.73 \text{ MW}; P_2 = 496.22 \text{ MW}; P_3 = 411.05 \text{ MW}; \lambda = 66.547 \text{ \$/MWh}$$

According to this unconstrained solution, units 1 and 3 operate within their limits, while the output of unit 2 exceeds its upper limit. Since this is not possible, we set P_2 at its upper limit $P_2 = 450$ MW. The net load that must be supplied by units 1 and 3 is then: $L_{net} = 1200 - 450 = 750$ MW. Performing an economic dispatch with these two units gives:

$$\lambda = 70.353 \text{ \$/MWh}$$

$$P_1 = 311.76 \text{ MW}; P_3 = 438.24 \text{ MW}; P_2 = 450 \text{ MW, with a total operating cost of}$$
$$\$60{,}451.47.$$

Plugging these values in the expressions for the marginal cost of these generating units we get:

$$\frac{dC_1(P_1)}{dP_1} = \frac{dC_3(P_3)}{dP_3} = \lambda = 70.353 \text{ \$/MWh while } \frac{dC_2(P_2)}{dP_2} = 61.00 \text{ \$/MWh.}$$

Because of the constraint, these units no longer operate at the same marginal cost. Producing more power with unit 2 and less with the other units would reduce the total cost but this is not possible because unit 2 is operating at its upper limit.

Example 6.6 *Effect of generator limits on economic dispatch* Let us revisit Example 6.4 considering the limits on the output of the generators. According to the unconstrained solution of that example, units 2 and 3 would operate within their limits. However, unit 1 would be operating below its lower limit. If we fix $P_1 = 200$ MW and recalculate the economic dispatch with only units 2 and 3 and a net load of 400 MW, we get:

$$\lambda = 33.769 \text{ \$/MWh}$$

$$P_2 = 223.08 \text{ MW}; P_3 = 176.92 \text{ MW}$$

$$P_1 = 200 \text{ MW}$$

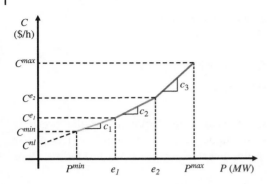

Figure 6.9 Piecewise-linear representation of the cost curve of a thermal generating unit.

Plugging these values in the expressions for the marginal cost of these generating units we get:

$$\frac{dC_2(P_2)}{dP_2} = \frac{dC_3(P_3)}{dP_3} = \lambda = 33.769 \text{ \$/MWh while } \frac{dC_1(P_1)}{dP_1} = 48.00 \text{ \$/MWh.}$$

The marginal cost of unit 1 is higher than the system marginal cost because it is at its lower limit and cannot produce any less.

6.8.3   Piecewise-Linear Cost Curves

Solving the economic dispatch problem using the method of Lagrangian multipliers is straightforward when the cost curves of the generating units are quadratic functions. However, in practice cost curves are typically modeled using piecewise linear functions. Figure 6.9. shows such a cost curves represented by a three-segment piecewise linear curve with two elbow points e_1 and e_2. The operating cost of a generating unit is then expressed as follows:

$$C = C^{nl} + c^1 P \quad \forall P \subset [P^{\min}, e_1]$$

$$C = C^{e_1} + c^2 (P - e_1) \quad \forall P \subset [e_1, e_2]$$

$$C = C^{e_2} + c^3 (P - e_2) \quad \forall P \subset [e_2, P^{\max}] \tag{6.34}$$

where:
c^1, c^2, c^3 are the slopes of the first, second, and third segments of the curve

$$C^{e_1} = C^{nl} + c^1 e_1$$

$$C^{e_2} = C^{e_1} + c^2 (e_2 - e_1)$$

Example 6.7   *Piecewise linear cost curves*   The table below shows a piecewise approximation of the quadratic cost curves of the generating units of Example 6.3. The elbow points have been chosen arbitrarily. Note that a piecewise representation of a cost curve is not necessarily less accurate than a quadratic representation if both are obtained by doing a regression on noisy measurement data.

	Unit 1	Unit 2	Unit 3
P^{min} (MW)	200	200	200
e_1 (MW)	300	300	300
e_2 (MW)	400	420	420
P^{max} (MW)	500	450	500
C^{min} ($/h)	7700	11,000	10,850
C^{e_1} ($/h)	13,500	14,700	15,250
C^{e_2} ($/h)	21,300	20,724	22,378
C^{max} ($/h)	31,100	22,500	28,250
c^1($/MWh)	58.00	37.00	44.00
c^2 ($/MWh)	78.00	50.20	59.40
c^3 ($/MWh)	98.00	59.20	73.40

When modeled using a piecewise-linear cost curve, the marginal cost of a generating unit (i.e., the derivative of its cost curve) is constant over each segment of its cost curve. Its marginal cost is therefore piecewise constant, as shown in Figure 6.10. This observation leads to the following computationally efficient economic dispatch algorithm.

1. Input the cost curve data of all the generating units
2. Create a list of all the segments of the cost curves of all the generators
3. Order this list in increasing order of marginal cost
4. Set the power output of all the generators at their respective minimum value
5. Set the residual load at the actual load minus the sum of these minimum generations
6. Take the first or next segment in the list
7. If the length of this segment is less than the residual load:
   a. Increase the output of the corresponding generator by the length of this segment
   b. Decrease the residual load by the length of this segment
   c. Go to step 6
8. Else
   a. Increase the output of the corresponding generator by the residual load
   b. Exit.

Figure 6.10 Piecewise-constant marginal cost curve of a thermal generating unit.

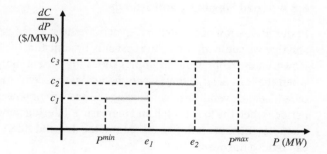

If you study this algorithm, you will notice that the power output of all the generators, except one, will be set at either P^{min}, P^{max}, or one of the elbow points of their cost curve. Only one generator operates somewhere along one of its segments. Since this generator would be the one to provide the next MW of load, it is called the marginal generator and the slope of this segment is the marginal system cost.

Example 6.8 *Economic dispatch with piecewise linear cost curves* Using the data of Example 6.7, determine the economic dispatch for loads of 800 and 1200 MW.

The table below shows an ordered list of the segments of the cost curves.

Segment number	1	2	3	4	5	6	7	8	9
Unit	2	3	2	1	2	3	3	1	1
Segment length	100	100	120	100	30	120	80	100	100
Marginal cost	37.0	44.0	50.2	58.0	59.2	59.4	73.4	78.0	98.0

We have $P_1^{min} + P_2^{min} + P_3^{min} = 600$ MW.

For an 800 MW load, the residual load is thus 200 MW, which is exactly covered by the first two segments. The economic dispatch is then $P_1 = 200.00$ MW; $P_2 = 300.00$ MW; $P_3 = 300.00$ MW. Since the next MW of load would be provided by the second unit on its second segment, the system marginal cost is 50.2 $/MWh. Using the data from Example 6.7, we can calculate the total operating cost:

$$C_{800} = C_1^{min} + C_2^{e_1} + C_3^{e_1} = 37,650 \text{ \$/h}$$

For a 1200 MW load, the residual load of 600 MW. To cover this residual load, we need to include all of the first six segments and 30 MW of the seventh. The economic dispatch is then $P_1 = 300.00$ MW; $P_2 = 450.00$ MW; $P_3 = 450.00$ MW. Since the system operates on the seventh segment, Unit 3 is the marginal unit, and the system marginal cost is 73.4 $/MWh. In this case, the total operating cost is:

$$C_{1200} = C_1^{e_1} + C_2^{max} + C_3^{e_2} + c_3^3 \times 30 = 60,580 \text{ \$/h}$$

6.8.4   Load Flexibility and Storage

In our discussion of economic dispatch, we have so far assumed that the load was a given and that we could minimize the cost only by adjusting the output of the various generators. However, as the previous examples show, the system marginal cost is significantly higher when the load is large than when the load is smaller. This suggests that we might be able to reduce the overall cost by reducing the amount of power produced by generators during periods when the load is high and increasing it during periods when it is low. The following examples illustrate how flexibility from the demand side, or arbitrage using energy storage, can reduce the overall cost.

Example 6.9 *Flexible load* Suppose that each of the two load levels that we considered in Example 6.8 occurs over separate one-hour periods. Let us calculate what the cost would be if one of the consumers were willing to shift 50 MW of its demand from the 1200 MW period to the 800 MW period.

Using the piecewise linear method, the economic dispatch for an 850 MW load is:

$$P_1 = 200.00 \text{ MW}; P_2 = 350.00 \text{ MW}; P_3 = 300.00 \text{ MW}$$

With a total operating cost:

$$C_{850} = C_1^{min} + C_2^{e_1} + c_2^2 \times 50 + C_3^{e_1} = 40{,}160 \text{ \$/h.}$$

For the 1150 MW load, the economic dispatch is:
$P_1 = 300.00 \text{ MW}; P_2 = 450.00 \text{ MW}; P_3 = 400.00 \text{ MW}$
With a total operating cost:

$$C_{1150} = C_1^{e_1} + C_2^{max} + C_3^{e_1} + c_3^2 \times 100 = 57{,}190 \text{ \$/h}$$

Adding the total operating costs over the two periods shows that shifting load from periods of high demand to periods of low demand reduces the overall cost:

$$C_{850} + C_{1150} = \$97{,}350 < C_{800} + C_{1200} = \$98{,}230$$

Example 6.10 *Storage* Instead of shifting 50 MW of demand from one hour to another, we could store 50 MWh in a battery during the low demand period and reduce the amount of energy that needs to be produced during the high demand period by discharging this stored energy.

The net load during the low demand period is $800 + 50 = 850$ MW. From Example 6.9, the cost of producing this power is:

$$C_{850} = 40{,}160 \text{ \$/h.}$$

However, we need to account for the losses incurred during the charging and discharging of the battery. Assuming an 80% roundtrip efficiency, the battery can only deliver during the high demand period:

$$E_{battery} = 0.8 \times 50 = 40 \text{ MWh}$$

The net load that must be supplied by the generators is then: $1200 - 40 = 1160$ MW. Using the data from Example 6.7, we find that the optimal economic dispatch is then:
$P_1 = 300.00 \text{ MW}; P_2 = 450.00 \text{ MW}; P_3 = 410 \text{ MW}$
With a total operating cost:

$$C_{1160} = C_1^{e_1} + C_2^{max} + C_3^{e_1} + c_3^2 \times 110 = 57{,}784 \text{ \$/h}$$

Summing the costs over the two periods, we find that:

$$C_{850} + C_{1160} = \$97{,}944 < C_{800} + C_{1200} = \$98{,}230$$

Performing energy arbitrage with a battery between these two periods is thus slightly profitable. A larger difference in system marginal cost between the discharging and charging periods would make it more profitable.

6.9 Unit Commitment

As we saw in Chapter 2, the load in a power system varies significantly over the course of the day. More generating units are clearly needed during periods when the load is high and fewer when it is low. In this section, we first use an example to develop an intuitive understanding of the factors that affect the optimal choice of generating units. We then formulate a mathematical technique to solve the unit commitment problem, which aims to select the units that should be used to optimally supply the load as it changes over the course of a day.

6.9.1 How Many Generating Units do we Need?

Example 6.11 *How many units?* The table below shows all the possible combinations of the generating units of Example 6.7 and how each of these combinations would be economically dispatched to supply a load of 475 MW.

Unit 1 (MW)	Unit 2 (MW)	Unit 3 (MW)	Total cost ($)	Marginal cost ($/MWh)
—	—	475	26,415.00	73.4
—	X	—	Infeasible	—
475	—	—	28,650.00	98.0
—	275	200	24,625.00	37.0
200	275	—	21,475.00	37.0
200	—	275	21,850.00	44.0
X	X	X	Infeasible	—

Supplying 475 MW with only Unit 2 is infeasible because this load is larger than $P_2^{\max}$. Supplying it with all three generating units is also impossible because $P_1^{\min} + P_2^{\min} + P_3^{\min} > 475$ MW. The other combinations are feasible but result in widely different total and marginal costs. These differences stem from the cost characteristics of these generating units. Looking back at the data from Example 6.3, we see that Unit 1 has a smaller no-load cost than the other two units, but that its operating cost increases more rapidly with its power output.

To explore in more detail the interplay between the no-load costs and the variable costs, let us perform economic dispatch calculations using the data from Example 6.5 for the entire range of loads that is feasible with these generating units. The dashed line in Figure 6.11 shows that the marginal cost increases monotonically with the load. The continuous line shows the average cost, which is defined as the total operating cost divided by the load:

$$C^{\mathrm{avg}} = \frac{C^{\mathrm{total}}(L)}{L} \tag{6.35}$$

The average cost curve factors in the no-load costs (i.e., the constant term in the cost functions), while the marginal cost curve does not because this constant term drops out when we take the derivative of the cost function. The average cost curve therefore exhibits

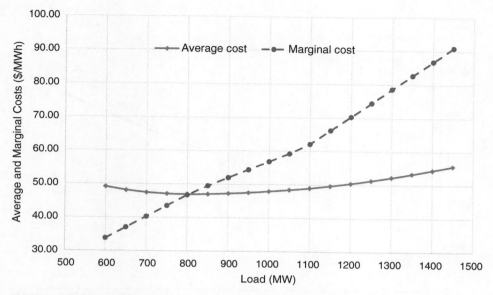

Figure 6.11 Average and marginal costs as a function of the load for the system of Example 6.5.

a minimum because the constant terms dominate for small values of the load. On the other hand, for large values of the load, the variable part of the cost, as reflected in the marginal cost, becomes prevalent. The existence of this minimum suggests that for small values of the load it might be desirable to reduce the fixed cost by operating fewer units or replacing some of them by units with a lower fixed cost and a higher variable cost. On the other hand, for large values of the load, operating more units at a lower marginal cost might be cost effective.

Figure 6.12 shows how the total cost varies as a function of the load when we perform an economic dispatch with combinations of 1, 2, or 3 of the generating units of Examples 6.7 and 6.8. Operating two generating units instead of three is cheaper in the 600–700 MW range. Using fewer units also extends the feasible range to lower loads. If the load were to exceed the sum of the P^{max} of the three units in these examples, we would need to commit more units.

6.9.2 Formulating the Unit Commitment Problem

Having argued that the optimal number and type of generating units varies with the load, let us turn our attention to how system operators might select these units. Each day, these operators forecast what the load is likely to be. They subtract from that forecast what they expect generation from renewable sources will provide to obtain the profile of net load that will have to be produced by the thermal generating units. Figure 6.13 shows an example of such a net load forecast. Typically, such forecasts are done with an hourly resolution, i.e., they assume that the load remains constant over each one-hour period.

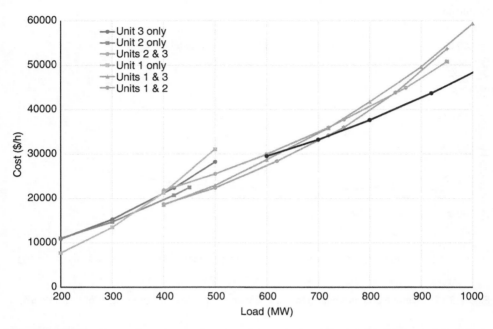

Figure 6.12 Total operating cost as a function of the load for different combinations of the units of Examples 6.7 and 6.8.

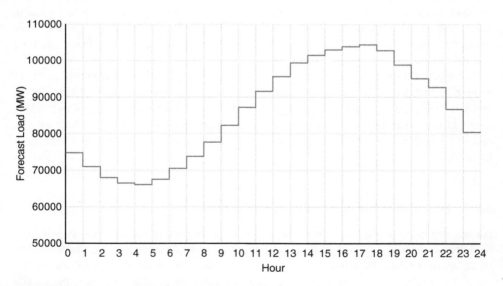

Figure 6.13 An example of a 24-hour hourly net load forecast.

Given this forecast and a set of available generating units, we could determine for each hour the subset of units that would supply the load at minimum cost. However, if we consider the whole day, these results would be neither optimal nor feasible. First, every time we start a thermal generating unit, a certain amount of fuel must be burned before it is ready to be synchronized and produce electric power. The cost of this fuel is called the startup cost C^{st} of the generating unit. It is incurred only at the time when the unit is brought online and should not be confused with the no-load cost C^{nl}, which is incurred each hour the unit is operating. The startup cost can be viewed as an investment. If a unit has a low variable cost but a large startup cost, it is worthwhile to bring it online only if this startup cost can be amortized by generating lots of cheap power over many hours. On the other hand, if we need to produce only a small amount of power during a short time, synchronizing a unit that is less efficient but has a larger variable cost may be the better option. Optimality thus requires that we schedule the generating units not on an hour-by-hour basis but with a horizon of at least one day. We therefore formulate the objective function of the unit commitment problem as follows:

$$\min_{u_i(t), P_i(t)} \sum_{i \in G} \sum_{t \in T} \left[C_i^{st}(t) + u_i(t).C_i(P_i(t)) \right] \tag{6.36}$$

where:

G is the set of available thermal generating units
T is the set of hourly time intervals included in the horizon
$u_i(t)$ is the status of unit i: $u_i(t) = 1$ if unit i is generating (i.e., "on") during period t and $u_i(t) = 0$ otherwise (i.e., "off")
$P_i(t)$ is the power produced by unit i during period t
$C_i^{st}(t)$ is the startup cost of unit i. It is non-zero only if unit i is turned on at time t, i.e., $u_i(t) = 1$ and $u_i(t-1) = 0$
$C_i(P_i(t))$ is the cost of producing P_i for an hour with unit i, as in Section 6.8.

At each hour t, the power produced by the generating units that are on at that hour must be equal to the load while operating within their limits:

$$\sum_{i \in G} u_i(t) \cdot P_i(t) = L(t) \quad \forall t \in T \tag{6.37}$$

$$P_i^{\min} \leq P_i(t) \leq P_i^{\max} \quad \forall i \in G, \forall t \in T \tag{6.38}$$

The second reason for looking ahead at least one day is that thermo-mechanical constraints limit the operational flexibility of these generating units. In particular, the amount by which a generating unit can increase or decrease its output between two subsequent time intervals is limited:

$$P_i(t) - P_i(t-1) \leq \Delta P_i^{up} \quad \forall i \in G, \forall t \in T \tag{6.39}$$

$$P_i(t-1) - P_i(t) \leq \Delta P_i^{down} \quad \forall i \in G, \forall t \in T \tag{6.40}$$

Since starting up a unit creates significant thermal gradient in its turbine, once turned on it should remain on for at least its minimum uptime $T_{up}^{\min}$ to allow temperatures to equalize:

$$\text{If } \left\{ u_i(t-1) = 1 \text{ and } \exists \, \tau > t - T_{up}^{\min} \text{ such that } u_i(\tau) = 0 \right\} \Rightarrow u_i(t) = 1 \tag{6.41}$$

Similarly, once a unit has been shut down, it must remain off for at least its minimum down time $T_{\text{down}}^{\min}$:

$$\text{If } \left\{ u_i(t-1) = 0 \text{ and } \exists\, \tau > t - T_{\text{down}}^{\min} \text{ such that } u_i(\tau) = 1 \right\} \Rightarrow u_i(t) = 0 \qquad (6.42)$$

Equations (6.39)–(6.42) thus create links between the various periods, forcing the optimization to extend over a sufficiently long time horizon.

6.9.3 Solving the Unit Commitment Problem

The unit commitment problem involves two types of decision variables:

- The binary decision variables $u_i(t)$, which specify the status of the generating units, i.e., which ones are "on," and which ones are "off" at every time interval t.
- The continuous decision variables $P_i(t)$, which specify how much power each of the units that are on produces at each time interval.

If we specify the status $u_i(t)$ of all the generating units at a particular time interval t, we can perform an economic dispatch with the units that are "on." In theory, to find the optimal solution to the unit commitment problem, we would have to consider all the possible combinations of the binary variables $u_i(t)$ at every time interval t, discard the ones that do not satisfy the constraints Eqs. (6.37)–(6.42), perform economic dispatches for all the other combinations over the horizon, and select the combination that result in the lowest total cost. However, with N generating units, there are 2^N possible combinations of unit status at each period. Since any combination at one period could be followed by any combination at the next period, finding the optimal schedule over T periods using this naïve approach would require considering $(2^N)^T$ combinations.

Example 6.12 *Number of combinations* Suppose that we need to schedule $N = 30$ generating units over $T = 24$ one-hour periods. In theory, to find the optimal schedule, we would have to consider $(2^{30})^{24} = 5.5 \times 10^{216}$ combinations.

If a computer were able to assess 1 billion combinations per second, it would need 1.7×10^{200} years to handle all of them.

Example 6.12 shows that considering all possible combinations is clearly impossible even in a medium-size system and even if many of these combinations can be immediately discarded because they don't satisfy the constraints. Fortunately, some clever mathematicians have developed "branch-and-cut" algorithms that identify a near-optimal solution in a reasonable amount of time. While these algorithms usually do not find the optimal solution, they provide an upper bound of the difference between the costs of a near-optimal solution and the actual optimal solution. If we model the cost curves of the generators using piecewise linear function as discussed in Section 6.8.3, the economic dispatch part of the problem can be solved using linear programming and the unit commitment problem becomes what is called a mixed integer linear programming problem. Using modern solvers, operators of large systems routinely schedule the commitment of hundreds of generating units over a few dozen periods.

Example 6.13 *Unit commitment* This example illustrates how various constraints affect the solution of the unit commitment problem. They were inspired by the documentation of the *pypsa* software and solved using this tool (Brown et al. 2018) and the glpk solver (https://www.gnu.org/software/glpk/). Interested readers can download this open-source software tool at https://pypsa.org/ and try modifications of these examples

Suppose that we have at our disposal two generating units to supply the following load profile:

Hour	1	2	3	4
Load (MW)	4000	800	5000	3000

We will start by assuming that the generating units have the following parameters:

Unit name	P^{min} (MW)	P^{max} (MW)	Marginal cost ($/MWh)
Hansel	3000	10,000	20
Gretel	500	5000	70

For simplicity, we assume that the cost functions of these generators have a single segment, and that their no-load and startup costs are zero.

Solving the unit commitment problem with this data produces the following solution:

Hour	1	2	3	4
P_{Hansel}	4000	0	5000	3000
P_{Gretel}	0	800	0	0

Since Hansel is much cheaper and has enough capacity, it carries the entire load during hours 1, 3, and 4. However, since the load at hour 2 is smaller than its P^{min}, it must be shut down and Gretel must be brought online.

Many generating units cannot be started for one hour and shut down immediately after. If we enforce a minimum uptime of three hours for Gretel, we get the following solution:

Hour	1	2	3	4
P_{Hansel}	3500	0	4500	3000
P_{Gretel}	500	800	500	0

Shutting down a large generating unit for only one hour is also unrealistic. Enforcing both a minimum uptime of three hours for Gretel and a minimum down time of two hours for Hansel leads to the following solution:

Hour	1	2	3	4
P_{Hansel}	4000	800	500	0
P_{Gretel}	0	0	4500	3000

Each added constraint not only changes the solution but also increases the operating cost:

	Cost ($)
Base case	296,000
Add min uptime	346,000
Add min down time	521,000

If we add a startup cost of $375,000 to Hansel, the optimization decides that it is cheaper not to start Hansel and to supply the entire load profile with Gretel even though the running cost of Gretel is considerably higher:

Hour	1	2	3	4
P_{Hansel}	0	0	0	0
P_{Gretel}	4000	800	5000	3000

6.10 Handling of Uncertainty

Our discussion of economic dispatch and unit commitment has so far assumed that we know precisely what the load and generation are going to be. However, in practice we must make allowance for several sources of uncertainty:

- On average, the best load forecasting models achieve errors slightly less than 2%. However, sometimes much larger errors occur, particularly during unusual weather conditions.
- Wind and solar generation depend directly on weather variables that can change very rapidly. For example, wind generation can drop or increase sharply when a weather front passes through an area with many wind farms. As the proportion of wind and solar generation in a system increases, this source of uncertainty becomes a significant concern.
- "Behind the meter" resources, such as residential solar generation, are particularly problematic because forecasters do not have access to data about their production.
- As we discussed in Section 6.7, conventional generating units occasionally fail at random times.

Power systems must therefore not only serve the load at minimum cost when things turn out as expected but also be able to cope when they don't. This means that they should have the resources needed to guard against the contingencies caused by these sources of uncertainty. Traditionally, these resources have been provided by the generation side and designed to cover a deficit in the load/generation balance. To ensure that sufficient resources are provided, we need to add a constraint to the formulation of the unit commitment. The headroom of a generating unit is the difference between the amount of power it is scheduled to produce and the maximum amount that it could produce:

$$P_i^{head}(t) = P_i^{max} - P_i(t) \tag{6.43}$$

At each time interval, the sum of the headroom of all the generating units must be greater than or equal to what is called the reserve requirement, which is an amount of power deemed sufficient to cope with any credible sudden imbalance in the load/generation balance:

$$\sum_{i \in G} u_i(t) P_i^{head}(t) = \sum_{i \in G} u_i(t) P_i^{max} - \sum_{i \in G} u_i(t) P_i(t) \geq R(t) \quad \forall t \in T \tag{6.44}$$

Combining this expression with Eq. (6.37), this reserve constraint takes the following form:

$$\sum_{i \in G} u_i(t) P_i^{max} \geq L(t) + R(t) \quad \forall t \in T \tag{6.45}$$

In small systems, the reserve requirement $R(t)$ is typically set at the capacity of the largest generating unit synchronized to the system because the loss of that unit is considered the largest credible contingency. In systems with a large number of generating units or with a substantial amount of wind or solar generation capacity, the reserve requirement is determined based on the probability distribution of load/generation imbalances. Since a sudden increase in renewable generation can cause a surplus of generation, systems must also have the ability to decrease the thermal generation. The overall rate at which generation can be increased or decreased must also be sufficient to avoid excessive frequency deviations. Including the reserve constraints forces the dispatch of some generating units to deviate from the economic optimum and often results in the commitment of additional units, both of which increase the cost of supplying the load. Providing reserve through voluntary load shedding or using energy storage devices reduces this undesirable effect.

Example 6.14 *Unit commitment with reserve* The tables below show the parameters of four generating units and the load profile that they must supply at minimum cost.

Unit name	pmin (MW)	pmax (MW)	Marginal cost ($/MWh)	Startup cost ($)	Initial state
Alpha	10	20	10	100	ON
Beta	5	30	12	100	ON
Gamma	5	50	13	100	OFF
Delta	10	20	20	0	OFF

Hour	1	2	3	4
Load (MW)	30	40	60	50

For simplicity, we assume that the cost functions of these generators have a single segment and that their no-load costs are zero. Alpha and Beta are the cheapest units and are already synchronized to the system at the beginning of the study period. Gamma is slightly more expensive and is initially offline. Delta has a high marginal cost but can be brought online at no cost. For simplicity, we do not consider minimum up- and down-time constraints.

If we do not impose a reserve constraint, the optimization tools mentioned in Example 6.13 provide the following unit commitment schedule:

Hour	1	2	3	4
P_{Alpha}	20	20	20	20
P_{Beta}	10	20	30	30
P_{Gamma}	0	0	0	0
P_{Delta}	0	0	10	0

Since Alpha is the cheapest, it produces at its P^{max} at all times. Beta, the next cheapest, produces the remainder, except at hour 3 when Delta is brought online to meet the load. While the marginal cost of Delta is much higher than that of Gamma, it does not carry a startup cost. The table below shows the resulting total hourly headroom.

Hour	1	2	3	4
Headroom	20	10	10	0

This schedule is thus unable to satisfy a 20 MW reserve requirement at hours 2, 3, and 4. Adding this reserve constraints to the optimization leads to the following optimal schedule:

Hour	1	2	3	4
P_{Alpha}	20	20	20	20
P_{Beta}	10	10	30	25
P_{Gamma}	0	10	10	5
P_{Delta}	0	0	0	0

Since Gamma must be brought online to satisfy the reserve requirement, turning on Delta just for hour 3 is no longer needed.

If in addition to the reserve requirement we enforce a maximum ramp up rate ΔP_i^{up} of 20 MW/h on Alpha, Beta, and Delta and of 5 MW/h on Gamma, the schedule changes again substantially:

Hour	1	2	3	4
P_{Alpha}	15	20	20	20
P_{Beta}	10	10	15	25
P_{Gamma}	5	10	15	5
P_{Delta}	0	0	10	0

As the table below shows, the operating cost increases each time we add a constraint.

	Cost ($)
Base case	2080
Add reserve constraint	2125
Add max ramp rate constraint	2225

Reference

Brown, T., Hörsch, J., and Schlachtberger, D. (2018). PyPSA: Python for power system analysis. *Journal of Open Research Software* 6 (1): arXiv:1707.09913. https://doi.org/10.5334/jors.188.

Further Reading

Kirschen, D.S. and Strbac, G. (2018). *Fundamentals of Power System Economics*, 2e. Wiley.
Wood, A.J., Wollenberg, B.F., and Sheblé, G.B. (2013). *Power Generation, Operation, and Control*, 3e. Wiley Interscience.

Problems

P6.1 At a given time the loads and losses in an isolated power system amount to 200 MW while the total generation is 204 MW. A battery energy storage system is used to maintain the power balance. How long can this situation last if the battery is initially fully discharged and has an energy capacity of 24 MWh?

P6.2 Once the battery of Problem P6.1 is fully charged, the generation is reduced to 190 MW while the load remains constant. How long can the battery maintain the power balance if its roundtrip efficiency is 85%?

P6.3 At what rate does the frequency decrease if the load suddenly exceeds the generation by 1000 MW in a large interconnected system where the $J\omega_m$ term in Eq. (6.9) is 20,000 MW/Hz²? How long would it take for the frequency to drop by 1.5 Hz if no control action occurs?

P6.4 Repeat problem P6.3 for an isolated system with a stiffness of 3000 MW/Hz².

P6.5 The frequency supplied by an isolated generator drops from 60 Hz to 57.0 Hz as the power it supplies increases from zero to its nominal value. What is the droop coefficient of its governor?

P6.6 A 60 Hz isolated generator rated at 50 kW has a droop coefficient of 4%. What frequency does it operate at when it supplies a load of 35 kW?

P6.7 Two 60 Hz generators operate in parallel to supply a load at a frequency of 58.5 Hz. The first generator is rated at 50 kW and the second at 75 kW. Both generators have a droop coefficient of 5%. What is the value of the load? How much power is each generator producing?

P6.8 A 60 Hz generator rated at 40 kW and a droop coefficient of 4% supplies a load of 30 kW. At what frequency does this generator operate? If a second generator rated at 30 kW and the same droop coefficient is connected in parallel, what is the new operating frequency? How much power does each generator provide?

P6.9 Generalize the expressions Eqs. (6.17)–(6.19) to the case of an arbitrary number of generators with the same droop coefficient.

P6.10 Generalize the expressions Eqs. (6.17)–(6.19) to the case of an arbitrary number of generators with different droop coefficients.

P6.11 A 15 MW generator with a droop coefficient of 0.04 and a 20 MW generator with a droop coefficient of 0.06 are connected in parallel and supply a load of 25 MW. If the nominal frequency is 60 Hz, calculate the operating frequency and the power produced by each generator. Hint: use the results of P6.10.

P6.12 At what frequency do underfrequency relays start shedding load in your region in case of a major system disturbance? How much load is shed at each step?

P6.13 The cost functions of the three generating units supplying a power system are as follows:

$$C_1(P_1) = 2000 + 6P_1 + 0.15\,P_1^2$$
$$C_2(P_2) = 7000 + 8P_2 + 0.12\,P_2^2$$
$$C_3(P_3) = 6000 + 7P_3 + 0.08\,P_3^2$$

For a load of 900 MW, calculate the economic dispatch, the system marginal cost, the marginal cost of each generating unit, and the total system operating cost.

P6.14 Repeat Problem P6.13 considering the following limits on the production of the generating units:

$$100 \leq P_1 \leq 300$$

$$200 \leq P_2 \leq 400$$

$$100 \leq P_3 \leq 350$$

P6.15 For what value of the load would the marginal cost of Problem P6.14 be equal to 85.00 $/MWh?

P6.16 Consider the generating units of Problems P6.13 and P6.14. Calculate the economic dispatch, the system marginal cost, the marginal cost of each generating unit, and the total system operating cost if 33.3% of the 900 MW is produced from wind and solar generation.

P6.17 A small power system is supplied by four generators. The cost characteristics of these generators are given by:

$$C_1(P_1) = 300 + 12P_1 + 0.05P_1^2$$

$$C_2(P_2) = 250 + 13P_2 + 0.06P_2^2$$

$$C_3(P_3) = 150 + 11P_3 + 0.08P_3^2$$

$$C_4(P_4) = 200 + 10P_4 + 0.07P_4^2$$

where the cost characteristic of each generator is expressed in units of $/h and the power provided by each generator is expressed in units of MW.

(a) Calculate the optimal economic dispatch for the case where the total load on the system is equal to 800 MW. Assume that there are no limits on the output of the generators.

(b) Assume that the load is unchanged at 800 MW but that that the following upper and lower limits are placed on the output of these generators:

$$50 \leq P1 \leq 200$$

$$70 \leq P2 \leq 230$$

$$100 \leq P3 \leq 300$$

$$240 \leq P4 \leq 350$$

Write the optimality conditions for this problem and solve it. What is the marginal cost of the equality and binding inequality constraints?

P6.18 Build piecewise linear cost curves for the units of Problems P6.13 and P6.14. For each unit use three segments of equal length. Display your results as in Example 6.7.

P6.19 Using the piecewise linear cost curves of Problem P6.18, calculate the economic dispatch, the system marginal cost, the marginal cost of each generating unit, and the total system operating cost for a load of 950 MW.

P6.20 Consider the following optimization problem:

$$minimize\, f(x, y) = 3x^2 - 12x + 4y^2 - 24y$$

subject to $x + y = 9.666$

$$7 \leq x \leq 24$$

$$0 \leq y \leq 8$$

(a) Solve this problem ignoring all constraints.
(b) Solve this problem ignoring the inequality constraints.
(c) Solve this problem considering all the constraints. Make sure to identify the binding constraints.

For each part, provide the optimal decision variables (primal variables), optimal objective function, and optimal Lagrange multipliers (dual variables), where appropriate (e.g., if the problem doesn't have constraints, there won't be any dual variables to consider).

P6.21 Four generating units are available to supply energy to a small power system. The characteristics of these units are summarized in Table P6.21A. On a given day, these units must be used to satisfy the load profile shown in Table P6.21B with a constant reserve requirement of 50 MW. Table P6.21C summarizes three solutions to this problem. Two of these are infeasible and one is feasible. Check the feasibility of each of these solutions. If a solution is infeasible, indicate all the constraints that are not satisfied. Indicate clearly which solution you believe is feasible.

Table P6.21A Generating u nit data.

Unit	p^{min}	p^{max}	Min uptime	Min down time	Initial status
A	200	500	6	6	Up for 12 h
B	60	150	3	3	Down for 1 h
C	150	250	3	3	Up for 6 h
D	50	100	1	1	Down for 6 h

Table P6.21B Load profile.

Hour	1	2	3	4
Load [MW]	500	600	700	400

Table P6.21C Solutions.

Solution	Units	Hour 1	Hour 2	Hour 3	Hour 4
S1	A	350	450	500	400
	B	OFF	OFF	OFF	OFF
	C	150	150	150	OFF
	D	OFF	OFF	50	OFF
S2	A	500	500	500	200
	B	OFF	OFF	OFF	OFF
	C	OFF	100	150	OFF
	D	OFF	OFF	100	200
S3	A	300	200	200	OFF
	B	OFF	100	200	OFF
	C	100	200	250	250
	D	100	100	100	150

P6.22 Table P6.22A gives the parameters of three generating units that can be used to supply the load profile given on Table P6.22B. Using the pypsa software (or any other suitable tool to which you have access), determine the minimum cost generation schedule and the value of the total operating cost for the following cases:

(a) Ignoring all constraints
(b) Adding to the previous case the maximum generation constraint on unit A
(c) Adding to the previous case the minimum generation constraint on unit B
(d) Adding to the previous case the minimum uptime constraint on unit B
(e) Adding to the previous case the minimum down time constraint on unit A
(f) Considering all the constraints of case e), what minimum value of startup cost on unit C would cause a change in the schedule?

Table P6.22A Generating unit data.

Unit name	p^{min} (MW)	p^{max} (MW)	Marginal cost ($/MWh)	Startup cost ($)	Min uptime (h)	Min down time (h)
A	60	100	20	0	1	3
B	25	50	25		4	1
C	10	40	50	0	3	1

Table P6.22B Load profile.

Hour	1	2	3	4	5
Load (MW)	80	110	120	80	70

7

Network Components

7.1 Overview

In the previous chapter, we implicitly assumed that electricity generation and consumption took place at the same location. However, since power systems usually connect generators and consumers that can be thousands of kilometers apart, we must study how power flows across networks. In this chapter we develop simple models of lines, transformers, and other components. In the following chapters, we will combine these models and use circuit analysis techniques to study the operation of large power systems. These circuit-based models are suitable for studying the steady-state and the dynamic behavior of power systems. More sophisticated models, which are not covered in this book, are required to study fast transients.

7.2 ac Lines

Because they are considerably cheaper than underground cables, three-phase overhead lines make up most of the branches of transmission networks and a substantial portion of the branches of distribution networks. However, underground cables are commonly used in urban areas either for aesthetic reasons or because they require less space. The structure of the models of both types of ac lines is identical, but the values of their parameters are quite different.

The resistance per unit of length of a conductor is given by:

$$r = \frac{\rho}{A_{\text{eff}}} \ (\Omega/m) \tag{7.1}$$

where ρ is the resistivity of the conducting material in Ωm and A_{eff} is the effective area of a cross section of the conductor in m^2. The effective area of a conductor is smaller than the geometric area of the cross section because ac currents are not uniformly distributed across the conductor.

Increasing the cross section of the conductor or constructing it using a lower resistivity material, e.g., copper instead of aluminum, reduces the I^2R ohmic losses. While this is desirable because it improves the energy efficiency of the power system, a balance must be struck between the benefits of reducing these losses and the cost of installing more

Power Systems: Fundamental Concepts and the Transition to Sustainability, First Edition. Daniel S. Kirschen.
© 2024 John Wiley & Sons Ltd. Published 2024 by John Wiley & Sons Ltd.
Companion website: www.wiley.com/go/kirschen/powersystems

expensive conductors. Ohmic losses also determine the amount of heat generated in the conductor and hence its temperature. In underground cables, excessive temperatures damage the insulation and will ultimately lead to a fault or short circuit. In overhead lines, they cause the conductor to dilate and sag. If a sagging line gets too close to a tree or other grounded object, a fault will occur. The maximum current that a conductor can safely carry is thus an essential parameter called its ampacity. To ensure reliability, system operators must monitor the flows in all branches of their network and check that it does not exceed their ampacity.

The currents flowing in an ac line create a magnetic field, which we model as a series inductance, whose value l per unit of length of the line is a function of the geometry of the line, i.e., the size and spacing of the conductors.[1] If we multiply the inductance per unit of length by the angular frequency, we get the series inductive reactance per unit of length of the line:

$$x_l = \omega l \; (\Omega/m) \tag{7.2}$$

As a rule of thumb, the series reactance of a transmission line is typically 10 times larger than its series resistance. For distribution lines, this ratio is closer to one.

Each phase of an extra-high-voltage (EHV) transmission line often consists of two or more conductors arranged in a bundle. Compared to a single conductor of equal current carrying capacity, such bundles reduce the inductance of the line.

The voltage applied to the transmission line creates an electric field, which we model as a capacitance connected between the conductor and ground (i.e., in a shunt connection). The value of this capacitance per unit of length of the line c is also determined by the geometry of the line. The shunt admittance per unit of length is then:

$$y_c = \omega c \; (S/m) \tag{7.3}$$

The shunt admittance of underground cables is significantly larger than that of overhead lines because the distance between the conductors is much smaller and the permittivity of the insulating material is higher than the permittivity of air.

A very strong electric field on the surface of a conductor creates the corona effect, which is an ionization of the air around the conductor. This effect causes radio interference, power losses, and a buzzing noise that you may have heard when passing under an EHV line. Bundling conductors reduces the intensity of the electric field and hence mitigates the corona effect. Losses associated with the corona effect is modeled by adding a shunt conductance g_c between conductor and ground.

We can therefore model a unit length of ac line by the circuit shown in Figure 7.1. As illustrated in Figure 7.2, to develop a rigorous model of an ac line, we must integrate the effects of these models over the length of the line. This integration leads to the π equivalent circuit shown in Figure 7.3. For lines shorter than a few hundred kilometers, multiplying

---

1 For a detailed derivation of the formulas used to calculate line parameters, the interested reader should refer to one of the books listed under Further Readings.

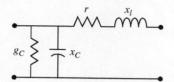

Figure 7.1 Model of a unit length of ac line.

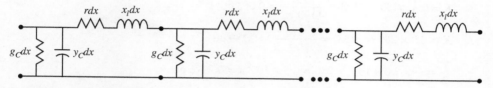

Figure 7.2 Rigorous model of an ac line.

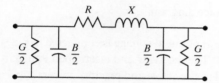

Figure 7.3 π equivalent circuit of an ac line.

the value of the parameters per unit of length by the length of the line L provides a good approximation of the parameters of this model:

$$R = r \times L$$
$$X = x_l \times L$$
$$B = y_c \times L$$
$$G = g_c \times L \tag{7.4}$$

The shunt conductance G is typically included only in models of EHV lines. The susceptance B is negligible for short lines ($L \leq 25$ km). In such cases, the model of the line simplifies to a resistance and inductive reactance in series.

Example 7.1 *Line parameters* A three-phase 345-kV line is 250 km long. Its parameters per unit of length are:

$$r = 0.3 \, \Omega/\text{km}$$

$$x_l = 0.9 \, \Omega/\text{km}$$

$$y_c = 2.0 \times 10^{-6} \, \text{S/km}$$

$$g_c \text{ is negligible}$$

Using a 100 MVA base, calculate the per unit parameters of its π-equivalent circuit.

From Table 4.1, we have:

$$Z_B = \frac{V_B^2}{S_B} = \frac{(345 \times 10^3)^2}{100 \times 10^6} = 1190.25\,\Omega$$

$$Y_B = \frac{1}{Z_B} = 840.16 \times 10^{-6}\,S$$

Hence:

$$R = 0.3 \times 250 = 75\,\Omega = \frac{75}{1190.25} = 0.063\,\text{pu}$$

$$X = 0.9 \times 250 = 225\,\Omega = \frac{225}{1190.25} = 0.189\,\text{pu}$$

$$\frac{Y}{2} = \frac{1}{2} \times 2.0 \times 10^{-6} \times 250 = 2.5 \times 10^{-4}S = 0.2976\,\text{pu}$$

$$X_c = \frac{1}{Y/2} = 3.361\,\text{pu}$$

Figure 7.4 summarizes these results in an impedance diagram.

Example 7.2 *Line model* A generator operating at nominal voltage is connected to the sending end of the transmission line of Example 7.1. The receiving end is left open-ended. Calculate the voltage $\overline{V}_R$ at the receiving end as well as the active and reactive powers supplied by this generator.

Figure 7.4 illustrates this example. The values of the shunt admittances have been converted to impedances. We can calculate $\overline{V}_R$ using the voltage divider equation:

$$\overline{V}_R = \frac{-j3.361}{0.063 + j0.189 - j3.361}\overline{V}_S = 1.0595\angle - 1.14°\,\text{pu} = 365.5\angle - 1.14°\,\text{kV}$$

The voltage at the receiving end is therefore higher than at the sending end. This phenomenon commonly occurs in lightly loaded or open-ended lines and is called the Ferranti effect.

The currents $\overline{I}_1$ and $\overline{I}_2$ are calculated as follows:

$$\overline{I}_1 = \frac{\overline{V}_S - \overline{V}_R}{R + jX} = \frac{1.0\angle0° - 1.0595\angle - 1.14°}{0.063 + j0.189} = 0.3152\angle88.86°\,\text{pu}$$

$$\overline{I}_2 = j\frac{Y}{2}\overline{V}_S = j0.2796 \times 1.0\angle0° = 0.2796\angle90°\,\text{pu}$$

The complex power supplied by the generator is then:

$$\overline{S} = P + jQ = \overline{V}_S(\overline{I}_1 + \overline{I}_2)^* = 1.0\angle0° \times 0.6128\angle - 89.41° = 0.00626 - j0.6127\,\text{pu}$$

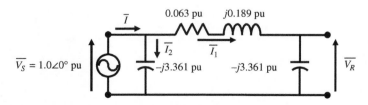

Figure 7.4 Impedance diagram of Examples 7.1 and 7.2.

Hence this generator produces 626 kW of active power, which is dissipated in losses in the line. To maintain its terminal voltage at 1.0 pu, it must also absorb 61.27 Mvar.

Example 7.3 *Line model* A load of 125 MW at 0.85 pf lagging is connected at the receiving end of the line of Example 7.1. If the receiving end voltage $\overline{V_R} = 1.0\angle0°$ pu, what is the voltage $\overline{V_S}$ at the sending end? How much active and reactive power does the generator produce?

In per unit, the active and reactive loads are:

$$P_L = \frac{125}{100} = 1.25 \text{ pu}$$

$$Q_L = \frac{125 \times \tan(\arccos(0.85))}{100} = 0.775 \text{ pu}$$

As illustrated in Figure 7.5, we can model this load as the parallel combination of a resistance R_L and an inductive reactance X_L:

$$R_L = \frac{V_R^2}{P_L} = \frac{(1.0)^2}{1.25} = 0.8 \text{ pu}$$

$$X_L = \frac{V_R^2}{Q_L} = \frac{(1.0)^2}{0.775} = 1.29 \text{ pu}$$

Since these impedances are in parallel with the line's shunt admittance, the current $\overline{I_1}$ is:

$$\overline{I_1} = \left(\frac{1}{0.8} + \frac{1}{j1.29} + \frac{1}{-j3.361}\right) \times 1.0\angle0° = 1.338\angle - 20.9° \text{ pu}$$

The voltage at the sending end is:

$$\overline{V_S} = \overline{V_R} + (R + jX)\overline{I_1} = 1.187\angle10.0° \text{ pu} = 409.54\angle10.0° \text{ kV}$$

The current $\overline{I_2}$ is:

$$\overline{I_2} = j\frac{Y}{2}\overline{V_S} = j0.2976 \times 1.187\angle10.0° = 0.353\angle100.0° \text{ pu}$$

Hence:

$$\overline{I} = \overline{I_1} + \overline{I_2} = 1.338\angle - 20.9° + 0.353\angle100.0° = 1.195\angle - 6.21° \text{ pu}$$

And the complex power supplied by the generator is:

$$\overline{S} = P + jQ = \overline{V_S}\overline{I}^* = 1.36 + j0.396 \text{ pu}$$

The generator thus supplies 125 MW to the load and 11 MW of line losses, as well as 39.6 Mvar.

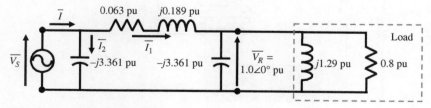

Figure 7.5 Impedance diagram of Example 7.3.

7.3 dc Lines

Each end of a dc line is connected to the rest of the system by a converter of the type described in Section 5.4.5. Since these converters precisely control the flow of power through the dc line, we can model dc lines as extracting a controlled amount of power at one end and injecting this same amount of power, minus losses, at the other.

7.4 Transformers

7.4.1 Single-Phase Transformer

As Figure 7.6 illustrates, a single-phase transformer consists of two windings wound around a core made of magnetic material. We arbitrarily designate one of these windings as the primary and the other as the secondary. If the primary winding loops around the core N_1 times, we say that it has N_1 turns. Similarly, the secondary winding has N_2 turns if it loops around the core N_2 times.

If we connect to the primary winding a time-varying voltage $e_1(t)$, according to Lenz's law, it creates a time-varying magnetic flux $\phi(t)$ such that:

$$e_1(t) = N_1 \frac{d\phi(t)}{dt} \tag{7.5}$$

Let us assume for now that this magnetic flux is perfectly contained in the magnetic core and links all the turns of both windings. Since it is time-varying, it induces a voltage in the secondary winding, which is again determined by Lenz's law:

$$e_2(t) = N_2 \frac{d\phi(t)}{dt} \tag{7.6}$$

Combining Eqs. (7.5) and (7.6), we get:

$$\frac{e_1(t)}{e_2(t)} = \frac{N_1}{N_2} \tag{7.7}$$

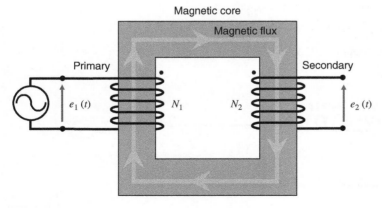

Figure 7.6 Components of a single-phase transformer.

The ratio between the primary and secondary voltages is therefore determined by the ratio of the number of turns in the primary and secondary windings.

If the voltage $e_1(t)$ is sinusoidal, the voltage $e_2(t)$ is therefore also sinusoidal and we can express Eq. (7.7) in terms of phasors:

$$\frac{\overline{E_1}}{\overline{E_2}} = \frac{N_1}{N_2} \tag{7.8}$$

Since the right-hand side of Eq. (7.8) is a real number, these two voltages are in phase.

As illustrated in Figure 7.7, let us connect an impedance Z_2 to the secondary winding so a current can flow in this winding. If we assume that there are no active or reactive power losses in this transformer, the complex power input on the primary side must be equal to the complex power output on the secondary side:

$$\overline{S_1} = \overline{S_2} \tag{7.9}$$

which expands as follows:

$$\overline{E_1}\,\overline{I_1}^* = \overline{E_2}\,\overline{I_2}^* \tag{7.10}$$

Combining Eqs. (7.8) and (7.10), we get:

$$\frac{\overline{I_1}}{\overline{I_2}} = \frac{N_2}{N_1} \tag{7.11}$$

The primary and secondary currents are thus in phase and the ratio of their magnitudes is equal to the inverse of the ratio of the number of turns in the primary and secondary windings.

The impedance Z_2 determines the ratio between the secondary voltage and the secondary current:

$$Z_2 = \frac{\overline{E_2}}{\overline{I_2}} \tag{7.12}$$

The ratio between the primary voltage and the primary current indicates how this impedance is "seen" from the primary side:

$$Z_2' = \frac{\overline{E_1}}{\overline{I_1}} \tag{7.13}$$

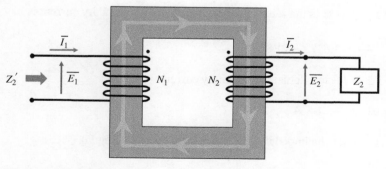

Figure 7.7 Single-phase transformer in the phasor domain.

Hence:

$$\frac{Z_2'}{Z_2} = \frac{\overline{E_1}/\overline{I_1}}{\overline{E_2}/\overline{I_2}} = \frac{\overline{E_1}\,\overline{I_2}}{\overline{E_2}\,\overline{I_1}} = \left(\frac{N_1}{N_2}\right)^2 \tag{7.14}$$

or:

$$Z_2' = \left(\frac{N_1}{N_2}\right)^2 Z_2 \tag{7.15}$$

When analyzing transformers, any impedance on the secondary side can therefore be replaced by an equivalent impedance on the primary side, calculated using Eq. (7.15).

Eqs. (7.8), (7.11), and (7.15) describe the behavior of an ideal single-phase transformer, i.e., a transformer where the core is made of a magnetic material with infinite permeability, where there are no losses, and where the same flux links all the turns of the primary and secondary windings. We can use these equations to get a first approximation of the voltages and currents in a transformer.

Example 7.4   The nameplate of a single-phase transformer indicates that its primary voltage rating is 240 V and its secondary voltage rating is 12 V. If the primary winding is connected to a 200 V source and the secondary winding to a 5+j2 Ω impedance, calculate the secondary voltage, the currents in the primary and secondary windings, and the active and reactive powers delivered by this transformer. Assume that the transformer is ideal, i.e., that the transformer itself consumes no active or reactive power.

We can infer the turns ratio of this transformer from the nominal voltages of the primary and secondary windings:

$$\frac{N_1}{N_2} = \frac{240}{12} = 20$$

The magnitude of the secondary winding voltage is then:

$$V_2 = \frac{N_2}{N_1}V_1 = \frac{1}{20} \times 200 = 10V$$

(Note that actual voltages do not have to be equal to the rated value. They can be lower but cannot be significantly higher because this could damage the insulation of the windings and cause a fault.)

Using the secondary voltage as the reference for the angles, the secondary current is:

$$\overline{I_2} = \frac{\overline{V_2}}{Z_2} = \frac{10\angle 0°}{5+j2} = 1.857\angle -21.8° A$$

We can then use Eq. (7.11) to calculate the primary current:

$$\overline{I_1} = \frac{N_2}{N_1}\overline{I_2} = 0.093\angle -21.8° A$$

Alternatively, we can refer the impedance Z_2 to the primary side using Eq. (7.15):

$$Z_2' = \left(\frac{N_1}{N_2}\right)^2 Z_2 = 400 \times (5+j2) = 2000 + j800\,\Omega$$

The primary current can then be calculated as follows:

$$\overline{I}_1 = \frac{\overline{V}_1}{Z_2'} = \frac{200}{2000 + j800} = 0.093\angle - 21.8°A$$

Since we assume that the transformer is ideal, the complex power is the same on both sides of this transformer:

$$S = P + jQ = \overline{V}_1 \overline{I}_1^* = \overline{V}_2 \overline{I}_2^* = 17.24 + j\,6.9\,VA$$

We need to expand the ideal transformer model to dispense with the simplifying assumptions and obtain results that match the behavior of actual transformers. As Figure 7.8 illustrates, the ideal transformer remains at the core of the model but is complemented by various components. The following paragraphs discuss these additions.

First, because the core is made of a magnetic material of high but not infinite permeability, a current is required to create the flux. Since this current exists whether or not a current flows in the secondary winding, we model it by adding a shunt reactance X_m.

Second, the flux creates eddy current and hysteresis losses in the magnetic core. To model these losses, we add a shunt resistance R_m.

Third, some of the flux that links the primary winding does not link the secondary winding and therefore does not contribute to the operation of the ideal transformer. Since this leakage flux is proportional to the current in the primary winding, we model it by adding a reactance X_1 in series with the primary winding. Similarly, we model the leakage flux that links the secondary winding but not the primary by introducing the series reactance X_2.

Finally, we model the ohmic losses in the primary and secondary windings by adding the series resistances R_1 and R_2.

The model of Figure 7.8 is clearly quite complex, particularly because it involves an ideal transformer which is not a standard component in circuit analysis. However, we are going to show that, if we work in the per unit system rather than in SI units, this ideal transformer disappears. As we discussed in Section 4.4, to express a voltage in per unit, we divide its value in Volts by the applicable base voltage. Therefore, the per unit voltages on the primary and secondary windings of an ideal transformer are:

$$E_1^{pu} = \frac{E_1}{E_1^{nom}} \tag{7.16}$$

$$E_2^{pu} = \frac{E_2}{E_2^{nom}} \tag{7.17}$$

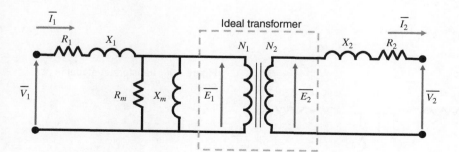

Figure 7.8 Transformer model in SI units.

where E_1^{nom} and E_2^{nom} are the nominal voltages for the primary and secondary sides of the transformer, respectively. Dividing Eq. (7.16) by (7.17), we get:

$$\frac{E_1^{pu}}{E_2^{pu}} = \frac{E_1}{E_2} \frac{E_2^{nom}}{E_1^{nom}} \tag{7.18}$$

Since in an ideal transformer we have $\frac{E_1}{E_2} = \frac{N_1}{N_2}$, if we are careful to choose:

$$\frac{E_1^{nom}}{E_2^{nom}} = \frac{N_1}{N_2} \tag{7.19}$$

Eq. (7.18) gives:

$$E_1^{pu} = E_2^{pu} \tag{7.20}$$

Similarly, the ratio of the per unit currents in the primary and secondary windings of an ideal transformer is:

$$\frac{I_1^{pu}}{I_2^{pu}} = \frac{I_1}{I_2} \frac{I_{2,B}}{I_{1,B}} \tag{7.21}$$

If we choose the base values for the voltages as above, the base values for the currents are:

$$I_{1,B} = \frac{S_B}{\sqrt{3} E_1^{nom}} \tag{7.22}$$

$$I_{2,B} = \frac{S_B}{\sqrt{3} E_2^{nom}} \tag{7.23}$$

Combining Eqs. (7.21–7.23) with (7.19), we get:

$$I_1^{pu} = I_2^{pu} \tag{7.24}$$

Equations (7.20) and (7.24) show that when we express all quantities in per units, an ideal voltage transformer has no effect on the voltages or currents. This observation leads us to the per unit form of the transformer model shown in Figure 7.9. We can further simplify this model because the magnitudes of the shunt impedances are usually much larger than those of the series impedances and thus have a negligible effect on the calculations. We can then combine the impedances of the primary and secondary windings as follows to obtain the model illustrated in Figure 7.10:

$$R_{eq}^{pu} = R_1^{pu} + R_2^{pu} \tag{7.25}$$

$$X_{eq}^{pu} = X_1^{pu} + X_2^{pu} \tag{7.26}$$

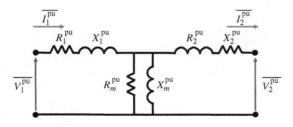

Figure 7.9 Transformer model in per units.

Figure 7.10 Simplified transformer model in per units.

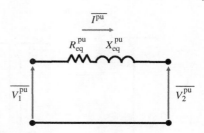

7.4.2 Three-Phase Transformer

We can construct a three-phase transformer by connecting together three single-phase transformers. Since the primary and secondary windings can be connected either in Y or in Delta, there are four basic transformer configurations: $Y-Y$, $Y-\Delta$, $\Delta-Y$, and $\Delta-\Delta$. When connected in Y, the neutral point can be connected to ground. Figure 7.11a shows the winding connections for the $Y-Y$ configuration. Figure 7.11b is a schematic representation of these windings intended to highlight the phase relation between the primary and secondary voltages. Figure 7.11c shows the corresponding phasor diagram. Finally, Figure 7.11d indicates how such a transformer would be represented on a one-line diagram. Figure 7.12 provides the same information for a $Y-\Delta$ configuration.

Rather than using three separate transformers, it is often cheaper to arrange the primary and secondary windings of all three phases on the same magnetic core, as illustrated in Figure 7.13. Note that in practice the primary and secondary windings are wound tightly together to reduce the leakage flux.

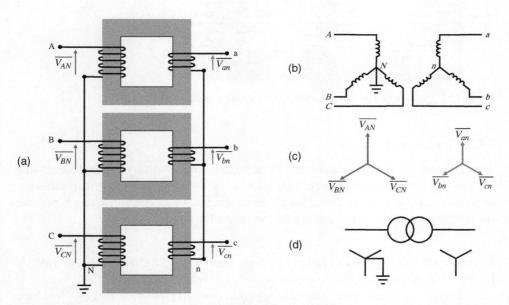

Figure 7.11 Three single-phase transformers connected in a Y-Y configuration and grounded on the primary side.

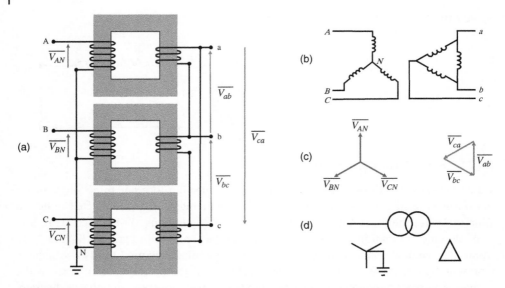

Figure 7.12 Three single-phase transformers connected in a Y-Δ configuration and grounded on the primary side.

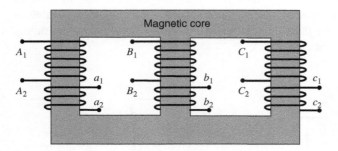

Figure 7.13 Schematic representation of a three-phase transformer.

7.4.3 Transformer Ratings

A transformer is characterized not only by its primary and secondary nominal line-to-line voltages but also by its MVA rating. This MVA rating is important because it specifies the maximum currents that can flow through the windings without creating so much loss that the resulting temperature rise might cause an insulation failure.

Example 7.5 *Transformer rating* A three-phase Y–Y 13.8 kV/416 V transformer is rated at 12 kVA. Calculate the maximum primary and secondary currents.

Recall that in a three-phase system, the apparent power is given by: $S = \sqrt{3}V_{LL}I_L$, where V_{LL} is the line-to-line voltage and I_L is the line current. The maximum line current on the primary side is therefore:

$$I_1^{max} = \frac{12000}{\sqrt{3} \times 13800} = 0.502\,A$$

while the maximum secondary line current is:

$$I_2^{max} = \frac{12000}{\sqrt{3} \times 416} = 16.65\,A$$

7.4.4 Three-Phase Transformers in the Per Unit System

If we adopt the per unit system that we introduced in Chapter 4, we can represent three-phase transformers by their single-phase equivalent, irrespective of the connection of the windings. As with other types of equipment, manufacturers provide the per unit parameters of a transformer on the basis of its rating. As we discussed in Section 4.4, when we incorporate a transformer in a system model, we must convert its per unit parameters to the common basis.

Example 7.6 *System parameters in per units* Figure 7.14a shows the one-line diagram of a small power system consisting of a generator, a step-up transformer, an ac line, a step-down transformer, and a load. The voltage at the load is at nominal value. The parameters of these components are given on the basis of their rating:

Generator: 75 MVA, 11 kV, $X_S = 1.83$ pu
Step-up transformer T_1: Δ–Y, 75 MVA, 11 kV/132 kV, $R_{T_1} = 0$, $X_{T_1} = 0.125$ pu
Line: 20 km long with $z = 0.18 + j0.40\,\Omega/km$
Step-down transformer T_2: Y–Δ, 45 MVA, 132 kV/33 kV, $R_{T_2} = 0$, $X_{T_2} = 0.125$ pu
Load: 10 MVA at 0.8 pf lagging

We first choose 100 MVA as the power basis for the entire system. We then choose a voltage basis for each zone: 11 KV for the generator and the low-voltage side of the step-up transformer, 132 kV for the line and the high-voltage side of the transformers, 33 kV for the load and the low-voltage side of the step-down transformer. These choices of voltage bases

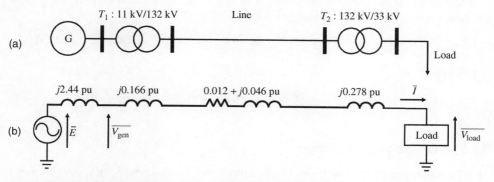

Figure 7.14 (a) One-line diagram of the small power system of Example 7.6; (b) Corresponding impedance diagram.

are compatible with the voltage ratings of the transformers. Given these choices, we can convert all the parameters to a consistent per unit basis using (4.37):

$$X_S^{new} = 1.83 \times \frac{100}{75} = 2.44 \text{ pu}$$

$$X_{T_1}^{new} = 0.125 \times \frac{100}{75} = 0.166 \text{ pu}$$

$$X_{T_2}^{new} = 0.125 \times \frac{100}{45} = 0.278 \text{ pu}$$

The impedance of the line is: $Z_{line} = 20 \times (0.18 + j0.40) = 3.6 + j8.0 \ \Omega$. To convert this impedance to per unit, we first calculate the base impedance using (4.41):

$$Z_B = \frac{(132 \times 10^3)^2}{100 \times 10^6} = 174.24 \ \Omega$$

Hence: $Z_{line} = \frac{3.6 + j8.0}{174.24} = 0.021 + j0.046$ pu

Converting the load to per units, we get:

$$\overline{S_{load}} = \frac{10}{100} \angle \arccos(0.8) = 0.1\angle 36.87° \text{ pu}$$

Figure 7.14b summarizes this data in a per unit impedance diagram.

If we take the voltage at the load as the reference for the angles, the current flowing through this system is:

$$\overline{I} = \left(\frac{\overline{S_{load}}}{\overline{V_{load}}} \right)^* = \left(\frac{0.1\angle 36.87°}{1.0\angle 0°} \right)^* = 0.1\angle -36.87° \text{ pu}$$

The voltage at the terminals of the generator is:

$$\overline{V_{gen}} = \overline{V_{load}} + \left(jX_{T_1}^{new} + Z_{line} + X_{T_2}^{new} \right)\overline{I}$$

$$\overline{V_{gen}} = 1.0\angle 0° + (j0.166 + 0.021 + j0.046 + j0.278)0.1\angle -36.87° = 1.032\angle 2.11° \text{ pu}$$

And the internal e.m.f. of the generator is:

$$\overline{E} = \overline{V_{gen}} + jX_S^{new}\overline{I} = 1.032\angle 2.1° + j2.44 \times 0.1\angle -36.87° = 1.2\angle 11.2° \text{ pu}$$

While the per unit current is the same across this system, the actual currents depend on the base current value applicable at each location:

Load: $I_B^{33kV} = \frac{S_B}{\sqrt{3}\,V_B} = \frac{100 \times 10^6}{\sqrt{3} \times 33 \times 10^3} = 1{,}749.5A \rightarrow \overline{I_{load}} = \overline{I} \times I_B^{33kV} = 174.95\angle -36.87°A$

Line: $I_B^{132\ kV} = \frac{S_B}{\sqrt{3}\,V_B} = \frac{100 \times 10^6}{\sqrt{3} \times 132 \times 10^3} = 437.4A \rightarrow \overline{I_{line}} = \overline{I} \times I_B^{132kV} = 43.74\angle -36.87°A$

Generator: $I_B^{11kV} = \frac{S_B}{\sqrt{3}\,V_B} = \frac{100 \times 10^6}{\sqrt{3} \times 11 \times 10^3} = 5248.7A \rightarrow \overline{I_{gen}} = \overline{I} \times I_B^{11kV} = 524.87\angle -36.87°A$

7.4.5 Tap-Changing Transformer

In a tap-changing transformer, the N_1/N_2 ratio is not fixed. Instead, as illustrated for a single-phase transformer in Figure 7.15, it depends on the mechanical position of a contactor that can tap at a limited number of positions on the secondary side. By changing

Figure 7.15 Schematic representation of a single-phase tap-changing transformer.

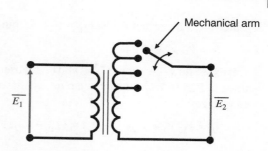

the tap position, we can therefore adjust the transformation ratio, and hence the secondary side voltage, in discrete steps. In an on-load tap-changing (OLTC) transformer, these adjustments can take place while the transformer is energized. Such transformers can therefore maintain the secondary voltage roughly constant as the voltage on the primary side varies.

As we saw in the previous section, the per unit model of a transformer is quite simple as long as the ratio of the number of turns is equal to the ratio of the base voltages on the primary and secondary sides of the transformer. Since in a tap-changing transformer the number of turns on the secondary is variable, this will often not be the case. We therefore need to develop a transformer model that accounts for a variable turns ratio.

If N_1 and N_2 are such that $N_1/N_2 = V_{1,B}/V_{2,B}$, the actual number of turns on the secondary N_2' is:

$$N_2' = tN_2 \tag{7.27}$$

If $t > 1$, the transformer boosts the voltage on the secondary side. It bucks that voltage if $t < 1$. The transformer is on its nominal tap ratio when $t = 1$. Converting voltages, currents, and impedances to per unit quantities eliminates the ideal transformer with turns ratio N_1/N_2 from the equivalent circuit of Figure 7.7. However, if the tap ratio $t \neq 1$, we must retain in our model an ideal transformer with a turn ratio $1 : t$ as shown in Figure 7.16. For simplicity, we neglect the shunt impedances of the transformer model of Figure 7.7 and combine all the series impedances in a single impedance Z on the tap-side (secondary) of the transformer.

Let us explore how we might be able to replace this model by one that does not involve an ideal transformer. If the voltage on the primary side of the ideal transformer is $\overline{V_1}$, the voltage on its secondary is $t\overline{V_1}$. Writing KVL on the secondary side, we have:

$$t\overline{V_1} - \overline{V_2} = Z\overline{I_2} \tag{7.28}$$

Expressing Eq. (7.28) in terms of the admittance $Y = 1/Z$, we have:

$$\overline{I_2} = tY\overline{V_1} - Y\overline{V_2} \tag{7.29}$$

Figure 7.16 Model of a tap-changing transformer. All quantities are in per unit.

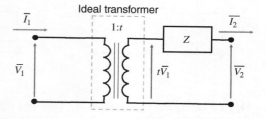

Since the transformer is ideal $\overline{I_1} = t\overline{I_2}$ and thus:

$$\overline{I_1} = t^2 Y \overline{V_1} - t Y \overline{V_2} \tag{7.30}$$

Let us see how Equations (7.29) and (7.30) translate into a π equivalent circuit of the form shown in Figure 7.17. In such a circuit, we have the following relations between currents and voltages:

$$\overline{I_1} = Y_{10}\,\overline{V_1} + Y_{12}(\overline{V_1} - \overline{V_2}) = (Y_{10} + Y_{12})\,\overline{V_1} - Y_{12}\overline{V_2} \tag{7.31}$$

$$\overline{I_2} = Y_{12}(\overline{V_1} - \overline{V_2}) - Y_{20}\,\overline{V_2} = Y_{12}\overline{V_1} - (Y_{12} + Y_{20})\,\overline{V_2} \tag{7.32}$$

Comparing Eqs. (7.29) and (7.30) with Eqs. (7.31) and (7.32), we conclude that the parameters of the circuit of Figure 7.17 must have the following values for this circuit to be equivalent to the model of Figure 7.16:

$$Y_{12} = tY$$
$$Y_{10} = t^2 Y - tY = t(t-1)Y$$
$$Y_{20} = Y - tY = (1-t)Y \tag{7.33}$$

Figure 7.18a illustrates this equivalent circuit. In Figure 7.18b, these admittances have been replaced by their equivalent impedances assuming that the series resistance of the transformer windings is negligible, i.e., $Z = 1/Y = jX$.

If the tap is on the neutral position (nominal tap ratio, $t = 1$), the shunt impedances in Figure 7.18b are infinite and the model is identical to the model of a transformer with a fixed turns ratio. If the tap increases the number of turns on the secondary side, i.e., $t > 1$, the factor $t(t-1)$ is positive while the factor $(1-t)$ is negative. The shunt element on the primary side of the equivalent circuit is then an inductance, while the one on the secondary side is a capacitance. Conversely, if $t < 1$, the factor $t(t-1)$ is negative, while the

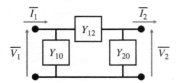

Figure 7.17 Generic π equivalent circuit.

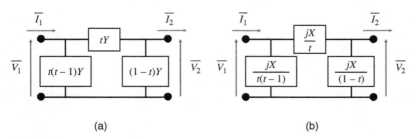

(a)　　　　　　　　　　(b)

Figure 7.18 π equivalent circuit of a tap changing transformer: (a) in terms of admittances; (b) in terms of impedances, assuming that the resistance of the windings is negligible.

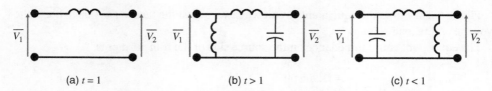

Figure 7.19 π equivalent circuit of a tap changing transformer: (a) nominal tap ratio; (b) higher than nominal tap ratio; (c) lower than nominal tap ratio.

factor $(1 - t)$ is positive. The shunt element on the primary side of the equivalent circuit is then a capacitance, while the one on the secondary side is an inductance. Figure 7.19 summarizes these observations. We can thus view changing the tap position as inserting a source of reactive power on one side and a sink of reactive power on the other.

Example 7.7 *Tap-changing transformer* As shown in Figure 7.20a, a load is connected to a source through a tap-changing transformer. The source maintains nominal voltage at its terminals and the load is modeled as a resistance $R_L = 1.0$ pu in parallel with an inductive reactance $X_L = 5.0$ pu When on its nominal tap position, the transformer is modeled as an inductive reactance $X_T = 0.1$ pu.

Figure 7.20b shows the equivalent circuit of this system when the taps of the transformer are on the nominal position $t = 1$. The voltage $\overline{V}_L$ at the load is:

$$\overline{V}_L = \frac{Z_L}{Z_L + jX_T}\overline{V}_S = \frac{0.962 + j0.192}{0.962 + j0.192 + j0.1}1.0\angle0° = 0.976\angle - 5.6° \text{ pu}$$

The current supplied by the source is:

$$\overline{I}_S = \frac{\overline{V}_S}{Z_L + jX_T} = 0.995\angle - 16.9° \text{ pu}$$

And the complex power supplied by the source is:

$$S_S = P_S + Q_S = \overline{V}_S \overline{I}_S^* = 0.952 + j0.289 \text{ pu}$$

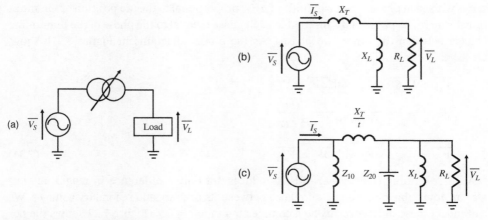

Figure 7.20 (a) System of Example 7.7; (b) Equivalent circuit when transformer taps are at nominal position $t = 1.0$; (c) Equivalent circuit when transformer taps are at position $t = 1.05$.

Figure 7.20b shows the equivalent circuit when we adjust the taps to $t = 1.05$ to boost the voltage on the load side.

The equivalent shunt impedance on the source side of the transformer is:

$$Z_{10} = \frac{1}{Y_{10}} = \frac{jX}{t(t-1)} = j1.905 \text{ pu}$$

The equivalent shunt impedance on the load side of the transformer is:

$$Z_{20} = \frac{1}{Y_{20}} = \frac{jX}{(1-t)} = -j2.0 \text{ pu}$$

The voltage $\overline{V}_L$ at the load is then:

$$\overline{V}_L = \frac{Z_L \| Z_{20}}{(Z_L \| Z_{20}) + j\frac{X_T}{t}} \overline{V}_S = \frac{0.917 - j0.275}{0.917 - j0.275 + j\frac{0.1}{1.05}} 1.0\angle 0° = 1.024\angle - 5.6° \text{ pu}$$

The source current is:

$$\overline{I}_S = \left(\frac{1}{(Z_L \| Z_{20}) + j\frac{X_T}{t}} + \frac{1}{Z_{10}} \right) \overline{V}_S = 1.097\angle - 16.9° \text{ pu}$$

And the complex power supplied by the source is:

$$S_S = P_S + Q_S = \overline{V}_S \overline{I}_S^* = 1.050 + j0.319 \text{ pu}$$

As expected, raising the taps increases the magnitude of the voltage on the load. It also increases the active and reactive power supplied by the source.

7.4.6 Phase-Shifting Transformer

Phase-shifting transformers introduce an adjustable phase angle difference between their primary and secondary sides. As Figure 7.21a illustrates, they typically consist of a series transformer and of a shunt transformer. The series transformer inserts a voltage $\overline{\Delta V}$ in series with each phase. The magnitude of $\overline{\Delta V}$ is proportional to the tap position of the shunt transformer. The phase of $\overline{\Delta V}$ inserted in each phase is equal to the phase of the line-to-line voltage between the other two phases. As the phasor diagrams in Figure 7.21b show, we have:

$$\overline{V}_{an} = \overline{V}_{AN} + \overline{\Delta V}_a \text{ with } \overline{\Delta V}_a \propto \overline{V}_{BC}$$

$$\overline{V}_{bn} = \overline{V}_{BN} + \overline{\Delta V}_b \text{ with } \overline{\Delta V}_b \propto \overline{V}_{CA}$$

$$\overline{V}_{cn} = \overline{V}_{CN} + \overline{\Delta V}_c \text{ with } \overline{\Delta V}_c \propto \overline{V}_{AB} \tag{7.34}$$

A phase-shifting transformer thus introduces not only a difference in magnitude but also an adjustable phase angle difference between its primary and secondary voltages. We can model these differences using a complex tap ratio $\bar{t} = t\angle\delta$. Figure 7.22 shows the per unit model of a phase-shifting transformer where we have neglected the shunt elements.

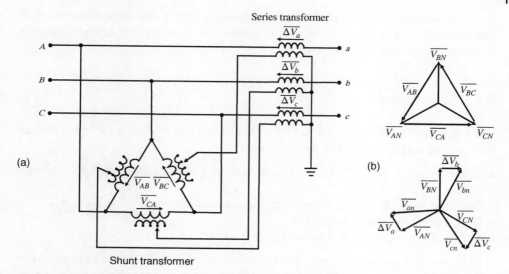

Figure 7.21 Schematic representation of a phase-shifting transformer.

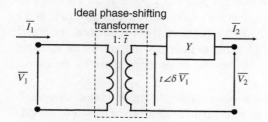

Figure 7.22 Model of a phase-shifting transformer. All quantities are in per unit.

Since the complex powers entering and exiting the ideal phase-shifting transformer must be equal, we have:

$$\overline{V}_1 \overline{I}_1^* = \overline{t}\,\overline{V}_1 \overline{I}_2^* \tag{7.35}$$

And thus:

$$\overline{I}_1 = \overline{t}^* \overline{I}_2 \tag{7.36}$$

Writing KVL on the secondary side, we have:

$$\overline{I}_2 = Y(\overline{t}\,\overline{V}_1 - \overline{V}_2) \tag{7.37}$$

Combining Eqs. (7.36) and (7.37), we get the equations that define the operation of a phase-shifting transformer:

$$\overline{I}_1 = t^2 Y \overline{V}_1 - \overline{t}^* Y \overline{V}_2$$

$$\overline{I}_1 = \overline{t} Y \overline{V}_1 - Y \overline{V}_2 \tag{7.38}$$

As we will see in the next chapter, using a phase-shifting transformer we can directly control the flow of active power in a transmission line. However, due to their high cost, such transformers remain relatively rare.

7.5 Switchgear

The topology of a power network can be altered by opening or closing switching devices. Such actions may be required for several reasons:

- Connecting a component to bring it into service and disconnecting it when it is no longer needed
- De-energizing and isolating a component for maintenance
- Disconnecting a faulted component
- Altering the power flows in a network.

Because power networks are generally inductive, interrupting a current can cause a large voltage difference $v = L\frac{di}{dt}$ across the contacts of the switch. A large voltage creates an arc that must be extinguished rapidly lest it damage or destroy the switch. Switching devices are therefore divided into circuit breakers and disconnect switches depending on whether they can or cannot interrupt a current. Circuit breakers are rated based on the magnitude of the current that they can interrupt. When specifying the installation of circuit breakers, power system designers must ensure that they will be able to interrupt the maximum expected fault or short circuit current, which is typically several times larger than the maximum load current.

Once a piece of equipment has been de-energized by opening of one or more circuit breakers, disconnect switches can be opened to provide further isolation, or closed to provide additional safety by connecting the equipment to ground.

Operators can open or close switching devices locally or remotely from a control center. Protective relays trigger the automatic opening of circuit breakers when they detect a short circuit in the network and determine that a piece of equipment must be disconnected to isolate the fault. In distribution networks, this type of switching action is often performed by fuses.

7.6 Reactive Compensation Devices

As we will discuss in the next chapter, reactive power has a direct effect on the voltages in a power network. Strategically located capacitors and inductors can help manage the balance of reactive power and hence the voltages. A shunt capacitor injects reactive power and raises the voltages in its electrical vicinity. Conversely, a shunt inductor extracts reactive power and lowers the voltages.

Switching capacitors and inductors in and out of service using circuit breakers alters the reactive injection or extraction in discrete steps and thus provides a coarse way of controlling the voltage. Connecting them using power electronics converters of the type described in Chapter 5 allows for a much finer voltage regulation.

7.7 Substations

Substation is the term used to designate the locations where transmission lines terminate and where distribution feeders originate. They also house transformers and larger reactive compensation devices. All these components are connected to busbars by switching devices. Figure 7.23 shows examples of possible substation designs. The choice of a particular configuration involves a tradeoff between flexibility and reliability on the one hand and cost on the other. For example, the straight bus configuration lacks flexibility because doing maintenance on any breaker requires the de-energization of a line or transformer. Moreover, a fault on the busbar puts the entire substation out of service. Such configurations are therefore typically only used at the interface between transmission and distribution networks where these issues are less pressing. The ring bus and breaker-and-a-half configurations are more flexible and reliable. However, they are more costly because they require more switchgear.

Busbars and closed switches have a negligible impedance, while open switches have an infinite impedance. All the components in a substation that are connected through closed switches are therefore tied to the same electrical node or bus. The number of buses at a particular substation is therefore a dynamic quantity that depends on the current on/off position of all the switching devices in that substation. For example, Figure 7.24 shows how opening three circuit breakers splits the breaker-and-a-half substation of Figure 7.23 into five buses.

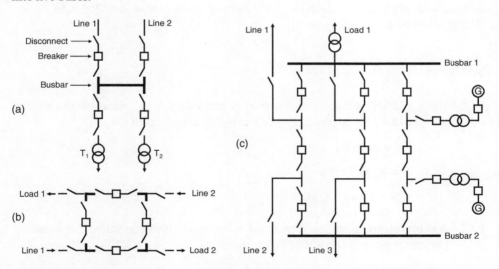

Figure 7.23 Examples of substation configurations: (a) straight bus; (b) ring bus; (c) breaker-and-a-half.

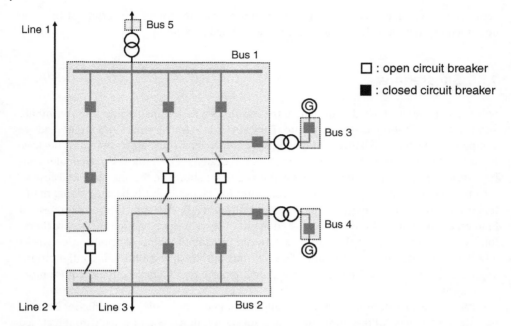

Figure 7.24 Example of demarcation of electrical nodes or buses in a substation.

Further Reading

Grainger, J. and Stevenson, W. (1994). *Power System Analysis*. McGraw Hill.
Glover, J.D., Overbye, T., and Sarma, M.S. (2016). *Power System Analysis and Design*, 6e. Cengage Learning.

Problems

P7.1 Using a 200 MVA basis, calculate the per unit values of the π-equivalent model of a 150 km long, three-phase, 230-kV transmission line, given that:

$$r = 0.04 \, \Omega/\text{km}$$

$$x_l = 0.5 \, \Omega/\text{km}$$

$$y_c = 1.5 \times 10^{-6} \, \text{S/km}$$

P7.2 A short transmission line is modeled using a simplified equivalent circuit whose per unit parameters on a 100 MVA basis are:

$$R = 0.05 \, \text{pu}$$

$$X = 0.25 \, \text{pu}$$

$$\frac{Y}{2} \text{ is negligible.}$$

A generator operating at its nominal voltage injects 80 MW and 50 Mvar at the sending end of this transmission line. Calculate the magnitude and phase of the receiving end voltage, the active and reactive power delivered at the receiving end, and the active and reactive losses in the line.

P7.3 A short transmission line delivers 50 MW at 0.95 pf lagging and nominal voltage on a 100 MVA basis. The power injected at the sending end of the line is 51.5 MW at 0.85 pf lagging. Calculate the per unit values of the line parameters if we model it as a resistance and reactance in series.

P7.4 The parameters of the π-equivalent circuit of a transmission line are:

$$R = 0.015 \, \text{pu}$$

$$X = 0.2 \, \text{pu}$$

$$\frac{Y}{2} = 0.5 \, \text{pu}$$

The shunt conductance is negligible.
When the receiving end of this line is left open-ended, the voltage at this receiving end is equal to its nominal value. Calculate:
(a) The magnitude and phase of the current flowing through the line
(b) The magnitude and phase of the voltage at the sending end of the line
(c) The magnitude and phase of the current entering the line
(d) The active and reactive power flowing into the line.

P7.5 Repeat the calculations of problem P7.4 for the case where this line supplies an active power load of 1.25 pu at 0.9 pf lagging and the voltage at the receiving end is again at its nominal value.

P7.6 A single-phase transformer is rated 15 kVA, 13 kV/220 V. What is the turns ratio of this transformer? What are the rated primary and secondary currents?

P7.7 An ideal single-phase transformer with $N_1 = 1500$ and $N_2 = 500$ is connected to an impedance Z_2 across its secondary winding. If $\overline{V_1} = 1000\angle0°$ V and $\overline{I_1} = 5\angle-30°\,A$, determine $\overline{V_2}, \overline{I_2}, Z_2$, and Z_2', i.e., the value of Z_2 referred to the primary side.

P7.8 A single-phase 100 kVA, 2400 V/240 V, 60 Hz transformer is used a step-down transformer. The load, connected to the secondary, draws 60 kW of active power and 30 kvar of reactive power at 230 V. Assuming the transformer is ideal. Calculate primary voltage and the load impedance referred to the primary.

P7.9 A single-phase practical transformer is rated at 13 MVA and 66 kV/11.5 kV. The shunt components of the equivalent circuit are negligible.
(a) Find the rated primary and secondary currents.

(b) When the secondary is shorted, the rated current flows when the voltage applied to the primary is 5.5 kV, and the measured power input is 100 kW. Determine R_{eq} and X_{eq} (referred to the primary).

P7.10 A single-phase 50 kVA, 2400 V/240 V transformer has an equivalent series impedance of $1+j2\Omega$ (referred to the primary). The shunt components of the equivalent circuit are negligible. The transformer delivers rated power at unity power factor to a load at the rated secondary voltage. Determine the real and reactive power delivered to the primary side of the transformer.

P7.11 Consider the single-phase system shown in Figure P7.11. The transformer nameplate data is as follows:

T1: 15 MVA, 13.8 kV/138 kV, $X_{eq} = 0.1$ pu
T2: 15 MVA, 138 kV/69 kV, $X_{eq} = 0.08$ pu
$R_{load} = 500\ \Omega$

All other impedances are assumed negligible. Draw the per unit impedance diagram of this system using a 15 MVA basis.

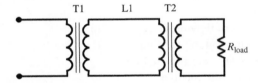

Figure P7.11 Figure for problem P7.11.

P7.12 Consider again the single-phase system shown in Figure P7.11. Draw the per unit diagram of this system using a 30 MVA basis given the following data about the various components:

T1: 20 MVA, 13.8 kV/138 kV, $X_{T1} = 0.1$ pu
T2: 15 MVA, 138 kV/69 kV, $X_{T2} = 0.08$ pu
L1: $Z_{line} = 2+j20\Omega$
$R_{load} = 300\ \Omega$

P7.13 The total series resistance and leakage reactance of a single-phase 132 kV–11 kV transformer referred to the primary side are 0.8 Ω and $j10$ Ω, respectively. This transformer supplies a load of 50 MW at 0.8 pf lagging and 10 kV on its low-voltage (secondary) side.
Calculate the current and voltage on the high-voltage side of this transformer.
Calculate the active power, reactive power, apparent power, and power factor on the primary side of this transformer. What is the efficiency of this transformer?

P7.14 Solve Problem P7.13 using per unit quantities using a 100 MVA base.

P7.15   Three of the transformers described in problem P7.6 are connected together to form a three-phase transformer bank. What are the primary voltage, secondary voltage, and MVA ratings of this transformer bank if these transformers are connected in $\Delta - \Delta$, $\Delta - Y$, $Y - \Delta$, and $Y - Y$?

P7.16   Consider the three-phase power system shown in Figure P7.16. The following data is provided about the various components:

Generator G1: 50 MVA, 13.2 kV, $x = 0.15$ pu
Generator G2: 20 MVA, 13.2 kV, $x = 0.1$ pu
Transformer T1: 80 MVA, 13.2/165 kV, $x = 0.08$ pu
Transformer T2: 40 MVA, 13.2/165 kV, $x = 0.1$ pu
L1: $50 + j200\ \Omega$
L2: $15 + j100\ \Omega$
L3: $20 + j50\ \Omega$
Load: $30 + j10$ MVA

Using a 100 MVA basis, draw an impedance diagram for this system showing the values of components in per unit.

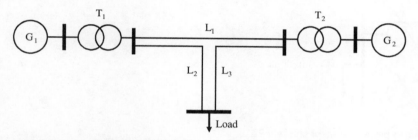

Figure P7.16   Small power system of problem P7.16.

P7.17   Figure P7.17 shows the one-line diagram of a small three-phase power system. The data about its components is as follows:

G: 90 MVA, 22 kV, $x = 0.18$ pu
T1: 50 MVA, 22/220 kV, $x = 0.1$ pu
T2: 40 MVA, 220/11 kV, $x = 0.06$ pu
T3: 40 MVA, 22/110 kV, $x = 0.064$ pu
T4: 40 MVA, 110/11 kV, $x = 0.08$ pu
L1: $z = j48.4\ \Omega$
L2: $z = j65.4\ \Omega$

The load is modeled as an impedance $Z_{\text{load}} = 1.5 + j1.0\ \Omega$.
Draw the impedance diagram for this system and calculate the power consumed by the load in per unit and SI units if the internal e.m.f. of the generator is 1.5 pu. Use a 100 MVA basis.

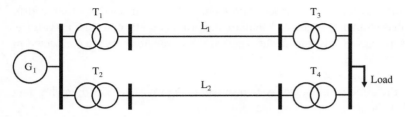

Figure P7.17 Small power system of problem P7.17.

P7.18 Consider the small system shown in Figure P7.18. A load of $0.8 + j0.6$ pu is supplied at nominal voltage by two tap-changing transformers connected in parallel. These two transformers are identical and have an equivalent series reactance $X = 0.1$ pu. Calculate the active and reactive power supplied by each transformer for the following two cases:

(a) The taps on both transformers are on the neutral position: $t_A = t_B = 1.0$
(b) The tap on transformer A is on the neutral position $t_A = 1.0$, while the tap on transformer B is $t_B = 1.05$
Discuss these results.

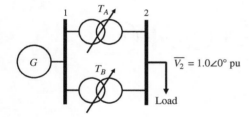

Figure P7.18 Small power system of problem P7.18.

P7.19 Consider the small system shown in Figure P7.19. A load of $0.9 + j0.6$ pu is supplied through the parallel combination of a line and a phase-shifting transformer. These components have the following parameters:

$L : R_L = 0; X_L = 0.2$ pu
$T : R_T = 0; X_T = 0.1$ pu
The shunt admittances are neglected.

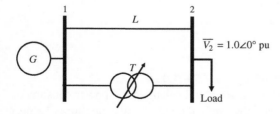

Figure P7.19 Small power system of problem P7.19.

Calculate the active and reactive power supplied through the line and the phase-shifting transformer for the following cases:

(a) The phase-shifting transformer is in the neutral position, i.e., it does not introduce a phase angle difference or a difference in voltage magnitude.
(b) The phase-shifting transformer introduces a voltage phase angle difference of +4 degrees without a difference in voltage magnitude.
(c) The phase-shifting transformer introduces a voltage phase angle difference of −4 degrees without a difference in voltage magnitude.

Discuss these results.

P7.20 Consider the substation one-line diagram of Figure P7.20. Assuming that all other switching devices are in the closed position, what components are connected together if the following breakers are in the open position?

(a) Breakers 2, 5, and 6 are open
(b) Breakers 3, 5, and 8 are open
(c) Breakers 9, 5, and 2 are open
(d) Breakers 1, 7, 5, and 3 are open.

Example: If breakers 2, 5, and 8 are open, we have two groups of components:
• Line 1, Load 1, and Generator G1 are connected together.
• Line 2, Line 3, and Generator G2 are connected together.

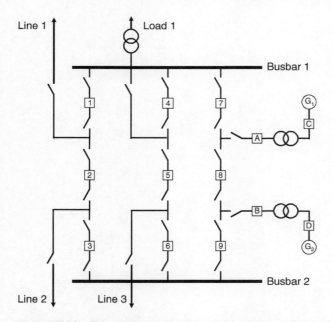

Figure P7.20 Substation one-line diagram for problem P7.20.

8

Power Flow

8.1 Overview

In this chapter, we combine the models of individual components that we developed in the previous chapter to construct a model of the overall network. Using this model, we will be able to calculate the steady-state voltages and flows across networks of any size or complexity. Given the load and generation at each node of the network, such a power flow calculation allows us to check that safe operating limits on flows and voltages will not be exceeded.

Our study of the power flow problem will proceed as follows:

- We will first develop a qualitative understanding of the relationships between the voltage magnitude and angle differences across a branch and the active and reactive power flows through that branch.
- We will then introduce nodal analysis, which is an efficient and systematic technique for assembling and solving equations representing large electrical networks.
- Building on nodal analysis, we will derive the power flow equations and a rigorous formulation of the power flow problem.
- Because the power flow equations are nonlinear, we will introduce the Newton–Raphson method, which is an iterative technique for solving this type of equations.
- We will then discuss applications of the power flow and the means of controlling flows and voltages.

8.2 Qualitative Relation Between Flows and Voltages

Consider, as illustrated in Figure 8.1, a line somewhere in the middle of a power system. This line connects bus 1 with voltage $\overline{V_1}$ to bus 2 with voltage $\overline{V_2}$. Assume that the complex power entering the line at bus 1 is $\overline{S} = P + jQ$. If we neglect the shunt elements of the equivalent circuit of the line, the current flowing through the line is:

$$\overline{I} = \frac{\overline{S}^*}{\overline{V_1}^*} = \frac{P - jQ}{V_1} \tag{8.1}$$

Power Systems: Fundamental Concepts and the Transition to Sustainability, First Edition. Daniel S. Kirschen.
© 2024 John Wiley & Sons Ltd. Published 2024 by John Wiley & Sons Ltd.
Companion website: www.wiley.com/go/kirschen/powersystems

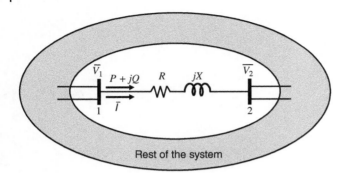

Figure 8.1 Small system to illustrate the relation between voltages and flows.

where we have taken $\overline{V}_1 = V_1 \angle 0°$ as the reference for the angles. The voltage at the other end of the line is then:

$$\overline{V}_2 = \overline{V}_1 - (R + jX)\overline{I} \tag{8.2}$$

Combining Eqs. (8.1). and (8.2) and separating the real and imaginary parts, we get:

$$\overline{V}_2 = \overline{V}_1 - \frac{RP + XQ}{V_1} - j\frac{XP - RQ}{V_1} \tag{8.3}$$

which we rewrite:

$$\overline{V}_2 = \overline{V}_1 - \Delta V_{Re} - j\Delta V_{Im} \tag{8.4}$$

In transmission networks, where the inductive reactance of lines is typically an order of magnitude larger than their resistance, we can make the following approximations:

$$\Delta V_{Re} \approx \frac{XQ}{V_1} \tag{8.5}$$

$$\Delta V_{Im} \approx \frac{XP}{V_1} \tag{8.6}$$

The phasor diagram of Figure 8.2 illustrates Eq. (8.4). If, as is usually the case, the phase angle difference between the voltages at the two ends of the line is reasonably small, this phasor diagram suggests that most of the difference in voltage magnitude stems from the ΔV_{Re} term, while the phase angle difference arises from the ΔV_{Im} term. Combining these geometrical observations with the approximations (8.5) and (8.6) leads us to the following assertions:

> The difference in voltage magnitude between the ends of a transmission line is mostly linked to the flow of reactive power through that line.
> The difference in voltage phase angle between the ends of a transmission line is mostly linked to the flow of active power through that line.

These assertions are not valid in distribution networks because the resistance and inductive reactance of a distribution line are typically of the same order of magnitude.

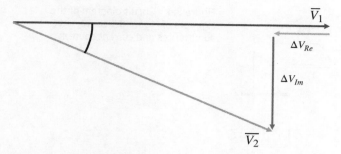

Figure 8.2 Phasor diagram illustrating the difference in voltages between the two ends of a line.

8.3 Nodal Analysis

All the examples we have considered so far involved networks with two or three nodes that could be solved using basic circuit analysis techniques. Because power networks often comprise thousands of nodes and branches, we need to introduce nodal analysis, which is a technique to automatically construct the network equations. We will explain the principle of nodal analysis using an example before generalizing the technique.

8.3.1 An Example of Nodal Analysis

Consider the five-node network shown in Figure 8.3. The first step in nodal analysis consists in converting each voltage source into its equivalent current source and replacing the branch impedances by their equivalent branch admittances:

$$\overline{I}_S = \frac{\overline{V}_S}{Z_S} \tag{8.7}$$

$$Y_x = \frac{1}{Z_x} \tag{8.8}$$

Figure 8.3 Small example used to illustrate the nodal analysis technique.

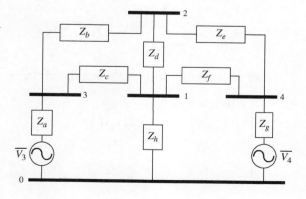

Figure 8.4 Circuit diagram of the example of Figure 8.3 converted to admittances and current sources.

Figure 8.4 shows the converted circuit diagram. Using the notations shown on this figure, let us write Kirchhoff's current law at each node, except at the reference node 0:

Node 1 : $\overline{I}_c + \overline{I}_d + \overline{I}_f + \overline{I}_h = 0$

Node 2 : $\overline{I}_b - \overline{I}_d + \overline{I}_e = 0$

Node 3 : $\overline{I}_a - \overline{I}_b - \overline{I}_c = \overline{I}_3$

Node 4 : $\overline{I}_g - \overline{I}_e - \overline{I}_f = \overline{I}_4$ $\qquad(8.9)$

We then express each branch current as a function of the nodal voltages. For example, $\overline{I}_c = Y_c(\overline{V}_1 - \overline{V}_3)$. Note that the voltage at the reference node 0 is zero. Since the currents supplied by the current sources are independent of the voltages, we put them on the right-hand side of the equations.

$$Y_c(\overline{V}_1 - \overline{V}_3) + Y_d(\overline{V}_1 - \overline{V}_2) + Y_f(\overline{V}_1 - \overline{V}_4) + Y_h(\overline{V}_1 - 0) = 0$$
$$Y_b(\overline{V}_2 - \overline{V}_3) - Y_d(\overline{V}_1 - \overline{V}_2) + Y_e(\overline{V}_2 - \overline{V}_4) = 0$$
$$Y_a(\overline{V}_3 - 0) - Y_b(\overline{V}_2 - \overline{V}_3) - Y_c(\overline{V}_1 - \overline{V}_3) = \overline{I}_3$$
$$Y_g(\overline{V}_4 - 0) - Y_e(\overline{V}_2 - \overline{V}_4) - Y_f(\overline{V}_1 - \overline{V}_4) = \overline{I}_4 \qquad(8.10)$$

In each of these equations, let us group together the factors of the various voltages:

$$(Y_c + Y_d + Y_f + Y_h)\overline{V}_1 - Y_d\overline{V}_2 - Y_c\overline{V}_3 - Y_f\overline{V}_4 = 0$$
$$- Y_d\overline{V}_1 + (Y_b + Y_d + Y_e)\overline{V}_2 - Y_b\overline{V}_3 - Y_e\overline{V}_4 = 0$$
$$- Y_c\overline{V}_1 - Y_b\overline{V}_2 + (Y_a + Y_b + Y_c)\overline{V}_3 = \overline{I}_3$$
$$- Y_f\overline{V}_1 - Y_e\overline{V}_2 + (Y_e + Y_f + Y_g)\overline{V}_4 = \overline{I}_4 \qquad(8.11)$$

Finally, let us write these equations in matrix form:

$$\begin{bmatrix} Y_c + Y_d + Y_f + Y_h & -Y_d & -Y_c & -Y_f \\ -Y_d & Y_b + Y_d + Y_e & -Y_b & -Y_e \\ -Y_c & -Y_b & Y_a + Y_b + Y_c & 0 \\ -Y_f & -Y_e & 0 & Y_e + Y_f + Y_g \end{bmatrix} \begin{bmatrix} \overline{V}_1 \\ \overline{V}_2 \\ \overline{V}_3 \\ \overline{V}_4 \end{bmatrix} = \begin{bmatrix} 0 \\ 0 \\ \overline{I}_3 \\ \overline{I}_4 \end{bmatrix} \qquad(8.12)$$

which we can write in shorthand as follows:

$$YV = I \tag{8.13}$$

This matrix equation expresses the vector of unknown nodal voltages V as a function of the vector of known source currents I and an admittance matrix Y.

The admittance matrix Y has a particular structure that applies to any network:

- Each row of Y corresponds to a node in the network because it stems from applying KCL at that node.
- Each column of Y also corresponds to a node in the network because it contains the factors of the voltage at that node.
- Diagonal element (i, i) of Y is the sum of all the admittances connected to node i.
- Off-diagonal element (i, j) of Y is equal to minus the admittance of the branch (or branches) connecting nodes i and j.
- Since element (i, j) is equal to element (j, i) for all nodes i and j, Y is a symmetric matrix.
- An admittance connecting a node to the reference node appears only in the diagonal element corresponding to that node.
- For a dc network the elements of Y are real numbers, while for an ac network they are complex numbers.

These rules can be translated easily into a computer program that can construct the admittance matrix, and hence the equations, for a network of arbitrary size and complexity.

Example 8.1 *Nodal equations* Consider the dc circuit diagram of Figure 8.5a. Figure 8.5b shows this same circuit with the impedances converted to admittances and the voltage sources converted to current sources. Using the node numbers shown in this figure, the nodal equations for this circuits are:

$$\begin{pmatrix} 1.083 & -0.333 & -0.25 \\ -0.333 & 0.7 & -0.166 \\ -0.25 & -0.166 & 1.416 \end{pmatrix} \begin{pmatrix} V_1 \\ V_2 \\ V_3 \end{pmatrix} = \begin{pmatrix} 5 \\ 0 \\ -10 \end{pmatrix}$$

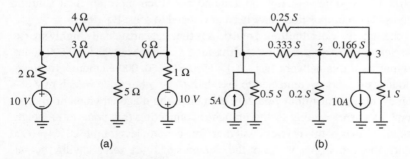

Figure 8.5 Circuit diagram for Example 8.1. (a) In terms of impedances and voltage sources; (b) in terms of admittances and current sources.

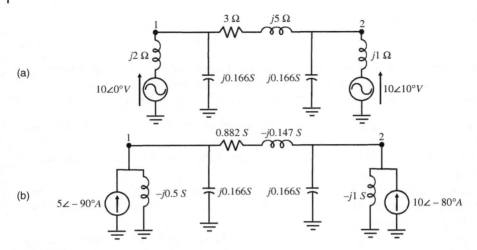

Figure 8.6 Circuit diagram for Example 8.2. (a) In terms of impedances and voltage sources; (b) in terms of admittances and current sources.

Example 8.2 *Nodal equations* Figure 8.6a shows the diagram of an ac circuit. Based on the admittances and current sources shown in Figure 8.6b, the nodal equations of this circuit are:

$$\begin{pmatrix} 0.882 - j0.481 & -0.882 + j0.147 \\ -0.882 + j0.147 & 0.882 - j0.981 \end{pmatrix} \begin{pmatrix} \overline{V_1} \\ \overline{V_2} \end{pmatrix} = \begin{pmatrix} 5\angle - 90° \\ 10\angle - 80° \end{pmatrix}$$

As long as there is at least one admittance connecting one of the nodes to the reference node, it is impossible to find a linear combination of rows or columns of Y that would add to a zero vector. The admittance matrix is thus nonsingular and invertible. The nodal voltages can therefore be expressed as follows:

$$V = Y^{-1}I \tag{8.14}$$

Solving Eqs. (8.1), (8.14) is not practical for large networks because inverting the admittance matrix Y is computationally too demanding. Much more efficient solution techniques, such as LU decomposition, have been developed to solve Eq. (8.13).

An important characteristic of admittance matrices is that they are usually highly sparse, i.e., most of their elements are zeros. To illustrate this fact, consider a network of 1500 nodes. The admittance matrix of this network has $1500 \times 1500 = 2,250,000$ elements. In power systems, each node is on average connected to less than three other nodes. Each row of Y thus contains on average at most four non-zero elements: the diagonal element and the off-diagonal elements corresponding to the branches connecting this node to its neighbors. The admittance matrix of this network will therefore contain less than $1500 \times 4 = 6000$ non-zero elements. In other words, 99.7% of the elements of Y are zero. Solving the network equations for large power systems would not be feasible without the use of special programming techniques that take advantage of this sparsity.

8.4 Formulation of the Power Flow Problem

Nodal analysis provides an efficient technique for writing and solving the equations of large electrical networks. However, we need to adapt it to model generators as injecting active power at a fixed voltage magnitude rather than as voltage or current sources. We must also model loads as extracting active and reactive power rather than as impedances. To this end, let us consider row k of the matrix Eq. (8.13):

$$\overline{I}_k = \sum_{i=1}^{N} Y_{ki}\overline{V}_i \tag{8.15}$$

This equation expresses Kirchhoff's current law at node k. The left-hand side represents the current that generators and loads inject at this node, while the right-hand side denotes the total power flowing from that node toward other nodes. If we take the complex conjugate of Eq. (8.15) and multiply it by the voltage at node k, the left-hand side becomes the complex power injected at node k:

$$\overline{S}_k = \overline{V}_k\overline{I}_k^* = \sum_{i=1}^{N} Y_{ki}^*\overline{V}_k\overline{V}_i^* \tag{8.16}$$

Let us adopt the following notations:

$$Y_{ki} = G_{ki} + jB_{ki} \tag{8.17}$$

$$\overline{V}_k = V_k\angle\theta_k \tag{8.18}$$

$$\overline{V}_i = V_i\angle\theta_i \tag{8.19}$$

Inserting these notations in the right-hand side of (8.16), we get:

$$\overline{S}_k = \sum_{i=1}^{N}(G_{ki} - jB_{ki})V_kV_i\angle(\theta_k - \theta_i) \tag{8.20}$$

Converting the voltage terms to rectangular coordinates yields:

$$\overline{S}_k = \sum_{i=1}^{N}(G_{ki} - jB_{ki})V_kV_i[cos\,(\theta_k - \theta_i) + jsin\,(\theta_k - \theta_i)] \tag{8.21}$$

Expanding the right-hand side and separating the real and imaginary parts, we obtain the power flow equations:

$$P_k^{inj} = Re\,(\overline{S}_k) = \sum_{i=1}^{N}V_kV_i[G_{ki}cos\,(\theta_k - \theta_i) + B_{ki}sin\,(\theta_k - \theta_{ii})] \tag{8.22}$$

$$Q_k^{inj} = Im\,(\overline{S}_k) = \sum_{i=1}^{N}V_kV_i\,[G_{ki}sin\,(\theta_k - \theta_i) - B_{ki}cos\,(\theta_k - \theta_i)] \tag{8.23}$$

where P_k^{inj} is the active power injected at node k. If the generators connected at that node inject P_k^G and the loads at that node extract P_k^L, we have:

$$P_k^{inj} = P_k^G - P_k^L \tag{8.24}$$

Similarly:

$$Q_k^{inj} = Q_k^G - Q_k^L \tag{8.25}$$

In a network with N nodes, we have N equations of the type (8.22) and N equations of the type (8.23). On the other hand, we have $4N$ variables: P_k^{inj}, Q_k^{inj}, V_k, and θ_k at each node k. To balance the number of equations and unknowns, two of these variables must be specified at each node. The type of bus determines which two variables are specified and which two are unknown.

PQ buses: At some buses, typically load buses, the active and reactive power injections P_k^{inj} and Q_k^{inj} are given and thus known. At these buses the voltage magnitude and angle V_k and θ_k are unknown.

PV buses: Generators inject a known amount of active power and maintain their terminal voltage at a constant values. The active power injection P_k^{inj} and the voltage magnitude V_k are therefore known at these buses, while the voltage angle θ_k and the reactive power injection Q_k^{inj} are unknown.

Slack bus: One bus in the system, usually a generator bus, must be designated as the slack or reference bus. At this bus, the voltage magnitude and angle are specified, while the active and reactive power injections are left to be determined. Two reasons compel the use of a slack bus. First, as in any ac circuit calculation, we need a reference for the angles, i.e., one bus where we set $\theta_k = 0°$. Second, when we solve a power flow problem, we assume that the system is in the steady state, which means that the power produced by the generators must be equal to the power consumed or dissipated in the system:

$$P_{generators} = P_{loads} + P_{losses} \tag{8.26}$$

If we fully specified the loads and generations at all buses when setting up a power flow problem, we would also have to know the losses in the system. However, these losses are a function of the voltages and flows, which we do not know because we have not yet solved this power flow. To resolve this contradiction, we leave the active power injection at the slack bus unspecified so that it can be adjusted to comply with Eq. (8.26).

These different bus types split the power flow Eqs. (8.22) and (8.23) into a set of implicit equations and a set of explicit equations. In the implicit equations, the left-hand side is known, while the right-hand side is a function of unknown variables. The set of implicit equations includes:

- Equations (8.22) and (8.23) for each PQ bus
- Equation (8.22) for each PV bus.

In the explicit equations, a single unknown variable is on the left-hand side. Its value can therefore be calculated once the values of the unknown variables on the right-hand side have been determined by solving the implicit equations. The set of explicit equations consists of:

- Equation (8.23) for each PV bus
- Equations (8.22) and (8.23) for the slack bus.

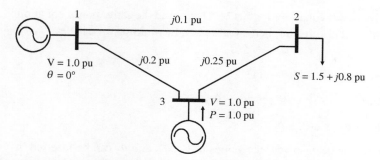

Figure 8.7 Three-bus power system of Example 8.3.

Example 8.3 *Power flow equation* Consider the three-bus power system shown in Figure 8.7. For simplicity, the lines are modeled as pure inductive reactances. After converting these line impedances to admittances, we can build the admittance matrix of this network:

$$Y = G + jB = j \begin{pmatrix} -15 & 10 & 5 \\ 10 & -14 & 4 \\ 5 & 4 & -9 \end{pmatrix}$$

Since all the G_{ki} terms are zero, expanding Eqs. (8.22) and (8.23) gives:

$$P_1^{inj} = -15V_1V_1\sin(\theta_1 - \theta_1) + 10V_1V_2\sin(\theta_1 - \theta_2) + 5V_1V_3\sin(\theta_1 - \theta_3)$$
$$Q_1^{inj} = +15V_1V_1\cos(\theta_1 - \theta_1) - 10V_1V_2\cos(\theta_1 - \theta_2) - 5V_1V_3\cos(\theta_1 - \theta_3)$$
$$P_2^{inj} = +10V_2V_1\sin(\theta_2 - \theta_1) - 14V_2V_2\sin(\theta_2 - \theta_2) + 4V_2V_3\sin(\theta_2 - \theta_3)$$
$$Q_2^{inj} = -10V_2V_1\cos(\theta_2 - \theta_1) + 14V_2V_2\cos(\theta_2 - \theta_2) - 4V_2V_3\cos(\theta_2 - \theta_3)$$
$$P_3^{inj} = +5V_3V_1\sin(\theta_3 - \theta_1) + 4V_3V_2\sin(\theta_3 - \theta_2) - 9V_3V_3\sin(\theta_3 - \theta_3)$$
$$Q_3^{inj} = -5V_3V_1\cos(\theta_3 - \theta_1) - 4V_3V_2\cos(\theta_3 - \theta_2) + 9V_3V_3\cos(\theta_3 - \theta_3)$$

which simplifies to the following if we observe that $\sin(0°) = 0$ and $\cos(0°) = 1$:

$$P_1^{inj} = 10V_1V_2\sin(\theta_1 - \theta_2) + 5V_1V_3\sin(\theta_1 - \theta_3)$$
$$Q_1^{inj} = +15V_1^2 - 10V_1V_2\cos(\theta_1 - \theta_2) - 5V_1V_3\cos(\theta_1 - \theta_3)$$
$$P_2^{inj} = +10V_2V_1\sin(\theta_2 - \theta_1) + 4V_2V_3\sin(\theta_2 - \theta_3)$$
$$Q_2^{inj} = -10V_2V_1\cos(\theta_2 - \theta_1) + 14V_2^2 - 4V_2V_3\cos(\theta_2 - \theta_3)$$
$$P_3^{inj} = +5V_3V_1\sin(\theta_3 - \theta_1) + 4V_3V_2\sin(\theta_3 - \theta_2)$$
$$Q_3^{inj} = -5V_3V_1\cos(\theta_3 - \theta_1) - 4V_3V_2\cos(\theta_3 - \theta_2) + 9V_3^2$$

Let us divide the buses into the three types:

- Bus 2 is a PQ bus: $P_2^{inj} = -1.5$ pu; $Q_2^{inj} = -0.8$ pu; V_2 and θ_2 are unknown
- Bus 3 is a PV bus: $P_3^{inj} = 1.0$ pu; $V_3 = 1.0$ pu; θ_3 and Q_3^{inj} are unknown
- Bus 1 is the slack bus: $V_1 = 1.0$ pu $\theta_1 = 0°$; P_1^{inj}; and Q_1^{inj} are unknown

The active and reactive power equations for bus 2 (PQ bus) and the active power equation for bus 3 (PV bus) form a set of three implicit equations in the unknown variables V_2, θ_2, and θ_3:

$$-1.5 = +10V_2 \sin(\theta_2) + 4V_2 \sin(\theta_2 - \theta_3)$$

$$-0.8 = -10V_2 \cos(\theta_2) + 14V_2^2 - 4V_2 \cos(\theta_2 - \theta_3)$$

$$+1.0 = +5 \sin(\theta_3) + 4V_2 \sin(\theta_3 - \theta_2)$$

Once these equations have been solved, the values of V_2, θ_2, and θ_3 can be inserted in the explicit active and reactive power equations for bus 1 (slack bus) and the reactive power equation for bus 3 (PV bus) to calculate P_1, Q_1 and Q_3:

$$P_1^{inj} = 10V_2 \sin(-\theta_2) + 5 \sin(-\theta_3)$$

$$Q_1^{inj} = +15 - 10V_2 \cos(-\theta_2) - 5 \cos(-\theta_3)$$

$$Q_3^{inj} = -5 \cos(\theta_3) - 4V_2 \cos(\theta_3 - \theta_2) + 9$$

8.5 Solving the Power Flow Equations

Eqs. (8.22) and (8.23) involve products of variables and transcendental functions. The power flow problem is thus non-linear and cannot be solved analytically or directly. Instead, an iterative technique, such as the Newton–Raphson method, is required. We will first explain the principle of this method using a single non-linear function of a single variable. Then, we will generalize its application to a system of non-linear equations. Finally, we will apply the Newton–Raphson method to the power flow problem.

8.5.1 Newton–Raphson Method

Consider a non-linear function $f(x)$ of a single variable x such as the one shown in Figure 8.8. We want to find a value x^* such that $f(x^*) = 0$. To apply the Newton–Raphson method, we must be able to calculate the value of the function $f(x)$ and the value of its derivative $f'(x)$ for any value of x. Suppose that we have reasons to believe that the solution x^* is somewhere in the vicinity of some value x^0. Unless we are very lucky, $f(x^0) \neq 0$. The value of $f'(x^0)$ gives us the slope of the tangent to $f(x)$ at $x = x^0$. As suggested by this figure, we can use these values to find x^1, which is closer to the solution:

$$x^1 = x^0 - \frac{f(x^0)}{f'(x^0)} \tag{8.27}$$

Unless the function $f(x)$ is linear, $f(x^1) \neq 0$ and we need to repeat the process using the following iteration formula until we find a value x^k such that we deem $f(x^k)$ sufficiently close to zero:

$$x^{k+1} = x^k - \frac{f(x^k)}{f'(x^k)} \tag{8.28}$$

Figure 8.8 Illustration of the principle of the Newton-Raphson method for a single function of a single variable.

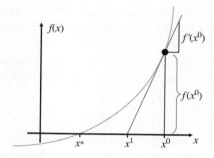

For a formal derivation of this iterative formula, recall that the Taylor series of an infinitely differentiable function is:

$$f(x) = f(x^0) + f'(x^0)(x - x^0) + \frac{1}{2}f''(x^0)(x - x^0)^2 + \dots \tag{8.29}$$

where f' denotes the first derivative of the function f, and f'' its second derivative.

If we neglect the terms of order two and higher, we get a linear approximation of the function around $x = x^0$:

$$f(x) \sim f(x^0) + f'(x^0)(x - x^0)$$

Eq. (8.27) then gives the value of x such that the value of this linearized approximation of $f(x)$ is equal to zero. Further iterations linearize the function around values of x that get iteratively closer to the solution.

Example 8.4 *One variable Newton–Raphson method* Let us use the Newton–Raphson method to find the solutions of the equation $f(x) \equiv x^2 - 12x + 32 = 0$. The table below shows how the method converges to the solution $x^* = 8$ when starting from $x^0 = 12$.

k	x^k	$f(x)$	$f'(x)$	x^{k+1}
0	12.000	32.000	12.000	9.333
1	9.333	7.111	6.667	8.267
2	8.267	1.138	4.533	8.016
3	8.016	0.063	4.031	8.000

If, on the other hand, we start from $x^0 = 2$, the method converges to the other solution $x^* = 4$.

k	x^k	$f(x)$	$f'(x)$	x^{k+1}
0	2.000	12.000	−8.000	3.500
1	3.500	2.250	−5.000	3.950
2	3.950	0.202	−4.100	3.999
3	3.999	0.002	−4.001	4.000

There is thus no guarantee that the method will converge to a particular solution. In some cases, it fails to converge or even diverge as is the case in this example if we choose $x^0 = 6$.

Let us see how we can generalize the Newton–Raphson to solve a system of two non-linear equations in two unknown variables:

$$\begin{cases} f(x, y) = 0 \\ g(x, y) = 0 \end{cases} \tag{8.30}$$

We can expand these two functions in a multivariable Taylor series:

$$f(x, y) = f(x^0, y^0) + \frac{\partial f(x^0, y^0)}{\partial x}(x - x^0) + \frac{\partial f(x^0, y^0)}{\partial y}(y - y^0) + \ldots$$

$$g(x, y) = g(x^0, y^0) + \frac{\partial g(x^0, y^0)}{\partial x}(x - x^0) + \frac{\partial g(x^0, y^0)}{\partial y}(y - y^0) + \ldots \tag{8.31}$$

If we neglect the higher order terms, we replace the solution of the system of non-linear Eq. (8.30) by the solution of the following system of linear equations:

$$\begin{cases} f(x^0, y^0) + \frac{\partial f(x^0, y^0)}{\partial x}(x - x^0) + \frac{\partial f(x^0, y^0)}{\partial y}(y - y^0) = 0 \\ g(x^0, y^0) + \frac{\partial g(x^0, y^0)}{\partial x}(x - x^0) + \frac{\partial g(x^0, y^0)}{\partial y}(y - y^0) = 0 \end{cases} \tag{8.32}$$

which can be rewritten as follows:

$$\begin{pmatrix} \dfrac{\partial f(x^0, y^0)}{\partial x} & \dfrac{\partial f(x^0, y^0)}{\partial y} \\ \dfrac{\partial g(x^0, y^0)}{\partial x} & \dfrac{\partial g(x^0, y^0)}{\partial y} \end{pmatrix} \begin{pmatrix} x - x^0 \\ y - y^0 \end{pmatrix} = - \begin{pmatrix} f(x^0, y^0) \\ g(x^0, y^0) \end{pmatrix} \tag{8.33}$$

The matrix of partial derivatives is called the Jacobian matrix $J(x, y)$ of the system of equations.

From Eq. (8.33) we can infer the two-dimensional equivalent of the iterative formula (8.28):

$$\begin{pmatrix} x^{k+1} \\ y^{k+1} \end{pmatrix} = \begin{pmatrix} x^k \\ y^k \end{pmatrix} - J(x^k, y^k)^{-1} \begin{pmatrix} f(x^k, y^k) \\ g(x^k, y^k) \end{pmatrix} \tag{8.34}$$

Example 8.5 *Two variables Newton–Raphson method* Let us solve the following system of two non-linear equations using the Newton–Raphson method:

$$\begin{cases} f(x, y) \equiv x + y - 15 = 0 \\ g(x, y) \equiv x \cdot y - 50 = 0 \end{cases}$$

The Jacobian matrix of this system of equation is:

$$J(x, y) = \begin{pmatrix} \dfrac{\partial f(x, y)}{\partial x} & \dfrac{\partial f(x, y)}{\partial y} \\ \dfrac{\partial g(x, y)}{\partial x} & \dfrac{\partial g(x, y)}{\partial y} \end{pmatrix} = \begin{pmatrix} 1 & 1 \\ y & x \end{pmatrix}$$

We arbitrarily choose $x^0 = 4$, $y^0 = 9$ as the starting point of our iterations. Inserting these values in the functions, we get:

$$f(x^0, y^0) = -2$$

$$g(x^0, y^0) = -14$$

$$J(x^0, y^0) = \begin{pmatrix} 1 & 1 \\ 9 & 4 \end{pmatrix}$$

$$J(x^0, y^0)^{-1} = \begin{pmatrix} -0.8 & 0.2 \\ 1.8 & -0.2 \end{pmatrix}$$

Inserting these values in (8.34) we get the values of the variables at the next iteration:

$$\begin{pmatrix} x^1 \\ y^1 \end{pmatrix} = \begin{pmatrix} 4 \\ 9 \end{pmatrix} - \begin{pmatrix} -0.8 & 0.2 \\ 1.8 & -0.2 \end{pmatrix} \begin{pmatrix} -2 \\ -14 \end{pmatrix} = \begin{pmatrix} 5.2 \\ 9.8 \end{pmatrix}$$

The table below summarizes the iterative process until convergence:

k	x^k	y^k	$f(x^k, y^k)$	$g(x^k, y^k)$	x^{k+1}	y^{k+1}
0	4.000	9.000	−2.000	−14.000	5.200	9.800
1	5.200	9.800	0.000	0.960	4.991	10.009
2	4.991	10.009	0.000	−0.045	5.000	10.000
3	5.000	10.000	0.000	0.000	5.000	10.000

8.5.2 Applying the Newton–Raphson Method to the Power Flow Problem

Consider a system of N buses. For simplicity, let us label the slack bus as bus 1 and the PV buses as buses 2 to m. Buses $m + 1$ to N are thus PQ buses. The following variables are known:

- θ_1: the voltage angle at the slack bus
- $V_1, \ldots, V_m$: the voltage magnitudes at the slack bus and the PV buses
- $P_2^{inj}, \ldots, P_N^{inj}$: the active power injections at the PV buses and PQ buses
- $Q_{m+1}^{inj}, \ldots, Q_N^{inj}$: the reactive power injections at the PQ buses.

On the other hand, the following variables are unknown:

- P_1^{inj}: the active power injection at the slack bus
- $Q_1^{inj}, \ldots, Q_m^{inj}$: the reactive power injections at the slack bus and the PV buses
- $\theta_2, \ldots, \theta_N$: the voltage angles at all buses except the slack bus
- $V_{m+1}, \ldots, V_N$: the voltage magnitudes at the PQ buses.

The first two sets of unknown variables appear explicitly on the left-hand side of Eqs. (8.22) or (8.23). We will therefore be able to calculate them directly from these equations once we have calculated the other unknown variables. These unknown voltage angles and magnitudes appear implicitly on the right-hand side of the Eq. (8.22) corresponding to the PV and PQ buses and of the Eq. (8.23) corresponding to the PQ buses.

To calculate these variables, we thus have to solve a system of $2N - 1 - m$ non-linear equations in $2N - 1 - m$ variables using the Newton–Raphson method. We can rewrite these equations as follows:

For $k = 2, \ldots, N$

$$P_k(\theta_2, \ldots, \theta_N, V_{m+1}, \ldots, V_N) \equiv \sum_{i=1}^{N} V_k V_i [G_{ki} \cos(\theta_k - \theta_i) + B_{ki} \sin(\theta_k - \theta_i)] - P_k^{inj} = 0$$

(8.35)

For $k = m + 1, \ldots, N$:

$$Q_k(\theta_2, \ldots, \theta_N, V_{m+1}, \ldots, V_N) \equiv \sum_{i=1}^{N} V_k V_i [G_{ki} \sin(\theta_k - \theta_i) - B_{ki} \cos(\theta_k - \theta_i)] - Q_k^{inj} = 0$$

(8.36)

The functions $P_k(\theta_2, \ldots, \theta_N, V_{m+1}, \ldots, V_N)$ and $Q_k(\theta_2, \ldots, \theta_N, V_{m+1}, \ldots, V_N)$ are called, respectively, the active and reactive mismatch functions because their values are the mismatch between the powers injected at a bus and the powers flowing from this bus toward other buses as calculated based on the angle and magnitude of the voltages at all buses. Each iteration of the Newton–Raphson method reduces these mismatches by adjusting these voltage magnitudes and angles and hence the branch flows. Equation (8.37) shows how to apply the Newton–Raphson method to the power flow problem:

$$\begin{pmatrix} \theta_2^{k+1} \\ \vdots \\ \theta_N^{k+1} \\ V_{m+1}^{k+1} \\ \vdots \\ V_N^{k+1} \end{pmatrix} = \begin{pmatrix} \theta_2^{k} \\ \vdots \\ \theta_N^{k} \\ V_{m+1}^{k} \\ \vdots \\ V_N^{k} \end{pmatrix} - J^{-1} \cdot \begin{pmatrix} P_2 \left(\theta_2^k, \ldots, \theta_N^k, V_{m+1}^k, \ldots, V_N^k \right) \\ \vdots \\ P_N \left(\theta_2^k, \ldots, \theta_N^k, V_{m+1}^k, \ldots, V_N^k \right) \\ Q_{m+1} \left(\theta_2^k, \ldots, \theta_N^k, V_{m+1}^k, \ldots, V_N^k \right) \\ \vdots \\ Q_N \left(\theta_2^k, \ldots, \theta_N^k, V_{m+1}^k, \ldots, V_N^k \right) \end{pmatrix}$$

(8.37)

where J^{-1} is the inverse of the Jacobian matrix of the mismatch equations:

$$J \left(\theta_2^k, \ldots, \theta_N^k, V_{m+1}^k, \ldots, V_N^k \right) = \begin{bmatrix} \dfrac{\partial P_2}{\partial \theta_2} & \cdots & \dfrac{\partial P_2}{\partial \theta_N} & \dfrac{\partial P_2}{\partial V_{m+1}} & \cdots & \dfrac{\partial P_2}{\partial V_N} \\ \vdots & \ddots & \vdots & \vdots & \ddots & \vdots \\ \dfrac{\partial P_N}{\partial \theta_2} & \cdots & \dfrac{\partial P_N}{\partial \theta_N} & \dfrac{\partial P_N}{\partial V_{m+1}} & \cdots & \dfrac{\partial P_N}{\partial V_N} \\ \dfrac{\partial Q_{m+1}}{\partial \theta_2} & \cdots & \dfrac{\partial Q_{m+1}}{\partial \theta_N} & \dfrac{\partial Q_{m+1}}{\partial V_{m+1}} & \cdots & \dfrac{\partial Q_{m+1}}{\partial V_N} \\ \vdots & \ddots & \vdots & \vdots & \ddots & \vdots \\ \dfrac{\partial Q_N}{\partial \theta_2} & \cdots & \dfrac{\partial Q_N}{\partial \theta_N} & \dfrac{\partial Q_N}{\partial V_{m+1}} & \cdots & \dfrac{\partial Q_N}{\partial V_N} \end{bmatrix}$$

(8.38)

At each iteration, the values of all these derivatives must be recalculated using the latest voltage magnitudes and angles. From the definition of the mismatch functions of Eqs. (8.35) and (8.36), we can derive the following expressions for the various terms of the Jacobian:

$\dfrac{\partial P_k}{\partial \theta_i}$ terms

For $i \neq k$: $\qquad \dfrac{\partial P_k}{\partial \theta_i} = V_k V_i [G_{ki} \sin(\theta_k - \theta_i) - B_{ki} \cos(\theta_k - \theta_i)]$

For $i = k$: $\qquad \dfrac{\partial P_k}{\partial \theta_k} = \displaystyle\sum_{\substack{i=1 \\ i \neq k}}^{N} V_k V_i [-G_{ki} \sin(\theta_k - \theta_i) + B_{ki} \cos(\theta_k - \theta_i)]$ $\qquad$ (8.39)

$\dfrac{\partial P_k}{\partial V_i}$ terms

For $i \neq k$: $\qquad \dfrac{\partial P_k}{\partial V_i} = V_k [G_{ki} \cos(\theta_k - \theta_i) + B_{ki} \sin(\theta_k - \theta_i)]$

For $i = k$: $\qquad \dfrac{\partial P_k}{\partial V_k} = \displaystyle\sum_{\substack{i=1 \\ i \neq k}}^{N} V_i [G_{ki} \cos(\theta_k - \theta_i) + B_{ki} \sin(\theta_k - \theta_i)] + 2G_{kk} V_k$ $\qquad$ (8.40)

$\dfrac{\partial Q_k}{\partial \theta_i}$ terms

For $i \neq k$: $\qquad \dfrac{\partial Q_k}{\partial \theta_i} = V_k V_i [-G_{ki} \cos(\theta_k - \theta_i) - B_{ki} \sin(\theta_k - \theta_i)]$

For $i = k$: $\qquad \dfrac{\partial Q_k}{\partial \theta_k} = \displaystyle\sum_{\substack{i=1 \\ i \neq k}}^{N} V_k V_i [G_{ki} \cos(\theta_k - \theta_i) + B_{ki} \sin(\theta_k - \theta_i)]$ $\qquad$ (8.41)

$\dfrac{\partial Q_k}{\partial V_i}$ terms

For $i \neq k$: $\qquad \dfrac{\partial Q_k}{\partial V_i} = V_k [G_{ki} \sin(\theta_k - \theta_i) - B_{ki} \cos(\theta_k - \theta_i)]$

For $i = k$: $\qquad \dfrac{\partial Q_k}{\partial V_k} = \displaystyle\sum_{\substack{i=1 \\ i \neq k}}^{N} V_i [G_{ki} \sin(\theta_k - \theta_i) - B_{ki} \cos(\theta_k - \theta_i)] + 2B_{kk} V_k$ $\qquad$ (8.42)

Example 8.6 *Newton–Raphson power flow* Let us compute the power flow in the two-bus power system shown in Figure 8.9.

Bus 1 is the slack bus. Its voltage magnitude and angle are given, but its net active and reactive injections are unknown. Bus 2 is a PQ bus. While its voltage magnitude and angle

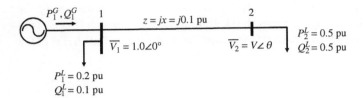

Figure 8.9 Two-bus power system of Example 8.6.

are unknown, its net active and reactive injections are given:

$$P_2^{inj} = -P_2^L = -0.5 \text{ pu}$$
$$Q_2^{inj} = -Q_2^L = -0.5 \text{ pu}$$

The admittance matrix of this system is:

$$Y = G + jB = j\begin{pmatrix} -10 & 10 \\ 10 & -10 \end{pmatrix} \text{ pu}$$

The power flow equations for bus 1 give explicitly the net active and reactive power injections as a function of the unknown voltage magnitude and angle:

$$P_1^{inj} = 10\,V \sin(-\theta)$$
$$Q_1^{inj} = 10 - 10V \cos(-\theta)$$

On the other hand, the power flow equations at bus 2 are implicit equations that we need to solve using the Newton–Raphson method. If we express them in mismatch form, we have:

$$P_2(\theta, V) \equiv 10\,V\,\sin\theta + 0.5 = 0$$
$$Q_2(\theta, V) \equiv 10\,V^2 - 10V\,\cos\theta + 0.5 = 0$$

The Jacobian matrix is:

$$J = \begin{pmatrix} \dfrac{\partial P_2}{\partial \theta} & \dfrac{\partial P_2}{\partial V} \\ \dfrac{\partial Q_2}{\partial \theta} & \dfrac{\partial Q_2}{\partial V} \end{pmatrix} = \begin{pmatrix} 10V\cos\theta & 10\sin\theta \\ 10\sin\theta & 20V - 10\cos\theta \end{pmatrix}$$

In the absence of other information, we start the iteration from what is called a "flat start," i.e., all the unknown voltages magnitudes are set at 1.0 pu and all the angles at zero degrees. Inserting $V^0 = 1.0$ pu and $\theta^0 = 0°$ in the above equations, the mismatches at bus 2 are:

$$P_2(\theta^0, V^0) = 0.5 \text{ pu}$$
$$Q_2(\theta^0, V^0) = 0.5 \text{ pu}$$

It is not surprising that these mismatches are exactly equal to the net injection at bus 2. With a flat start, no power flows in any line because all the buses are at the same voltage. To proceed with the first iteration, we need to calculate the Jacobian and its inverse:

$$J = \begin{pmatrix} 10 & 0 \\ 0 & 10 \end{pmatrix} \text{ and } J^{-1} = \begin{pmatrix} 0.1 & 0 \\ 0 & 0.1 \end{pmatrix}$$

$$\begin{pmatrix} \theta^1 \\ V^1 \end{pmatrix} = \begin{pmatrix} \theta^0 \\ V^0 \end{pmatrix} - J^{-1} \begin{pmatrix} P_2(\theta^0, V^0) \\ Q_2(\theta^0, V^0) \end{pmatrix} = \begin{pmatrix} 0 \\ 1.0 \end{pmatrix} - \begin{pmatrix} 0.1 & 0 \\ 0 & 0.1 \end{pmatrix} \begin{pmatrix} 0.5 \\ 0.5 \end{pmatrix} = \begin{pmatrix} -0.05 \text{ radians} \\ 0.95 \text{ pu} \end{pmatrix}$$

After this first iteration, the voltage at bus 2 lags the voltage at bus 1 and its magnitude is smaller. These adjustments induce a flow of active and reactive power from bus 1 to bus 2 and reduce the mismatches:

$$P_2(\theta^1, V^1) = 0.0252$$

$$Q_2(\theta^1, V^1) = 0.0369$$

Since these mismatches are greater than our tolerance of 0.001 pu, we continue iterating:

$$J = \begin{pmatrix} 9.488 & -0.4998 \\ -0.4748 & 10.0125 \end{pmatrix} \text{ and } J^{-1} = \begin{pmatrix} 0.10566 & 0.005274 \\ 0.00501 & 0.100125 \end{pmatrix}$$

$$\begin{pmatrix} \theta^2 \\ V^2 \end{pmatrix} = \begin{pmatrix} \theta^1 \\ V^1 \end{pmatrix} - J^{-1} \begin{pmatrix} P_2(\theta^1, V^1) \\ Q_2(\theta^1, V^1) \end{pmatrix} = \begin{pmatrix} -0.05 \\ 0.95 \end{pmatrix} - \begin{pmatrix} 0.10566 & 0.005274 \\ 0.00501 & 0.100125 \end{pmatrix} \begin{pmatrix} 0.0252 \\ 0.0369 \end{pmatrix}$$

$$= \begin{pmatrix} -0.0528 \text{ radians} \\ 0.9462 \text{ pu} \end{pmatrix}$$

After this iteration, the mismatches are:

$$P_2(\theta^2, V^2) = 0.00011$$

$$Q_2(\theta^2, V^2) = 0.003997$$

The active mismatch is within tolerance, but the reactive mismatch is still too large. Meeting this convergence criterion requires two more iterations, which produce the following solution:

$$\theta = -0.0529 \text{ radians} = -3.03°$$

$$V = 0.9458 \text{ pu}$$

Inserting these values in the explicit power flow equations for bus 1, we get:

$$P_1^{inj} = 0.5 \text{ pu}$$

$$Q_1^{inj} = 0.554 \text{ pu}$$

Since there is a load of $0.2 + j\,0.1$ pu at bus 1, the active and reactive power injected by the generator at that bus are:

$$P_1^G = 0.7 \text{ pu}$$

$$Q_1^G = 0.654 \text{ pu}$$

No active power is lost in this system because our line model does not include a resistance. On the other hand, the inductance of the line absorbs $0.554 - 0.5 = 0.054$ pu of reactive power.

8.6 Calculating the Line Flows

Once we have obtained the voltage magnitude and angle at each bus in the system by solving the power flow equations, we can calculate the flows in each branch of the network. Figure 8.10 illustrates the notations that we will use to derive the formulas. Figure 8.10a (which is similar to Figure 7.3) shows the π-model of the branch between nodes i and k. In Figure 8.10b the series and shunt components of this model have been replaced by equivalent admittances, which are calculated as follows:

$$Y_{ik}^{se} = \frac{1}{R_{ij} + jX_{ij}} = \frac{R_{ij}}{R_{ij}^2 + X_{ij}^2} - j\frac{X_{ij}}{R_{ij}^2 + X_{ij}^2} \tag{8.43}$$

$$Y_{ik}^{sh} = \frac{G_{ij}}{2} + j\frac{B_{ij}}{2} \tag{8.44}$$

The current entering line ik at node i is equal to the sum of the current in the series admittance between nodes i and k, and the current in the shunt admittance at node i:

$$\overline{I}_{ik} = Y_{ik}^{se}(\overline{V}_i - \overline{V}_k) + Y_{ik}^{sh}\overline{V}_i \tag{8.45}$$

Similarly, the current entering the line at node k is:

$$\overline{I}_{ki} = Y_{ik}^{se}(\overline{V}_k - \overline{V}_i) + Y_{ik}^{sh}\overline{V}_k \tag{8.46}$$

If the shunt components of the branch model have a negligible admittance, we have:

$$\overline{I}_{ik} = Y_{ik}^{se}(\overline{V}_i - \overline{V}_k) = -\overline{I}_{ki} \tag{8.47}$$

The complex power entering the line at node i can then be calculated as follows:

$$\overline{S}_{ik} = P_{ik} + jQ_{ik} = \overline{V}_i \overline{I}_{ik}^* = Y_{ik}^{se*}\left(|V_i|^2 - \overline{V}_i \overline{V}_k^*\right) + Y_{ik}^{sh*}|V_i|^2 \tag{8.48}$$

Similarly, the complex power entering the line at node k is:

$$\overline{S}_{ki} = P_{ki} + jQ_{ki} = \overline{V}_k \overline{I}_{ki}^* = Y_{ik}^{se*}\left(|V_k|^2 - \overline{V}_k \overline{V}_i^*\right) + Y_{ik}^{sh*}|V_k|^2 \tag{8.49}$$

P_{ik} and P_{ki} always have opposite signs because active power enters the branch at one end and exits at the other, with the losses accounting for the difference between their

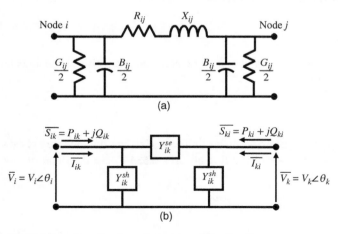

Figure 8.10 Notations for the calculations of the line flows.

magnitudes. While Q_{ik} and Q_{ki} usually have also opposite signs, they can both be positive if the branch consumes a lot of reactive power, or both negative if the branch produces more reactive power than it consumes.

As mentioned in Section 7.2, system operators must monitor the flows in all branches of their network to check that they do not exceed their ampacity and cause an excessive sag or an insulation failure. The apparent power flowing in a line is often used as a proxy for the current because the voltages typically do not deviate very much from their nominal value.

$$S_{ik} = |\overline{S_{ik}}| = V_i I_{ik} \text{ and } S_{ki} = |\overline{S_{ki}}| = V_k I_{ki} \tag{8.50}$$

Each branch in a network thus has a rated MVA value $S_{ik}^{\max}$. The flow in branch $i - k$ should therefore be such that:

$$S_{ik} \leq S_{ik}^{\max} \text{ and } S_{ki} \leq S_{ik}^{\max} \tag{8.51}$$

8.7 Power Flow Applications

Engineers compute power flows to determine the voltage and flows that would result under various postulated conditions. Besides the parameters of all network components, which do not change unless new network components are built and commissioned, the following data must be predicted or specified for each case:

- The load at every node.
- The active power produced by each generator connected to the network.
- The voltage at the terminals of each of these generators.
- The lines, transformers, and other components that are in service.

Long-term planners use the results of these computations to assess the value of building different transmission lines or installing reactive compensation devices. Medium-term planners perform power flow calculations to assess how taking some components out for maintenance would affect the operation of the system. Short-term planners and operators rely on power flows to identify unplanned line or generator outages that would lead to unacceptable operating conditions. Power flow computations can be automated to analyze a wide range of conditions and report only those that result in violations of an operating limit. In particular, contingency analysis is the term used to describe the systematic application of the power flow to a predetermined list of contingences to identify the problematic ones.

As we will see in later chapters, power flow computations also provide the initial conditions for fault calculations and dynamic time-domain simulations.

Example 8.7 *Power flow on a 5-bus system* Figure 8.11 shows the one-line diagram of a 5-bus power system. Solar farms are connected at buses 2 and 3, and a windfarm is connected at bus 4. Since the thermal-generating plant at bus 1 provides the power required to keep the system in balance, we have chosen this bus as the slack bus. When needed, an expensive thermal-generating plant can also be connected to bus 2. A reactive power compensation device is available at bus 5. The acceptable range for voltages is 0.95 pu $\leq V \leq 1.05$ pu. The parameters of the lines are given on a 100 MVA base in Table 8.1 and can also be downloaded from https://github.com/Power-Systems-Textbook/TextbookSimulations.

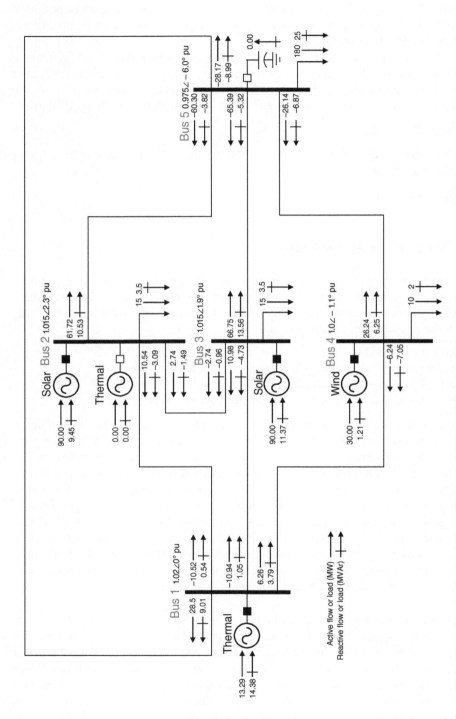

Figure 8.11 Power flow for the conditions of Example 8.7 (i.e., midday, all lines in service).

Table 8.1 Parameters of the 5-bus power system.

From bus	To bus	R (pu)	X (pu)	B (pu)	S^{max} (MVA)
1	2	0.01938	0.3917	0.0288	
1	3	0.03403	0.3040	0.0392	80
1	5	0.03699	0.3791	0.0338	75
1	4	0.02699	0.3397	0.0342	
2	3	0.02567	0.2632	0.024	
2	5	0.03695	0.2388	0.0246	90
3	5	0.03010	0.2103	0.0128	95
4	5	0.01335	0.3211	0.031	95

Figure 8.11 illustrates the results of a power flow computation for typical conditions around midday. A negative value indicates a flow in the direction opposite to that of the arrow.

The load is moderately high, the solar farms are producing at rated output, and there is also some significant wind generation. Observe the resulting pattern of active power flows and how it affects the voltage angles. Since the load is relatively light, maintaining the voltage at bus 5 within normal operating limits does not require the energization of the reactive power compensation device.

Example 8.8 *Correcting a line overload* Figure 8.12 shows how the disconnection of line 3–5 would affect the power flows for the midday conditions described in the previous example. This contingency would cause a mild violation of the lower voltage limit at bus 5. More significantly, it would result in an overload of line 2–5, with $S_{2-5} = \sqrt{P_{2-5}^2 + Q_{2-5}^2} = 98.95$ MVA greater than $S_{2-5}^{max} = 90$ MVA. As Figure 8.13 shows, curtailing the solar generation at bus 2 to 60 MW and compensating by increasing the generation at bus 1 reduces S_{2-5} to 89.47 MVA, which resolves the issues.

Example 8.9 *Correcting another overload* Figure 8.14 shows what the power flows would be in that same system for the evening peak load if no action were taken. After sunset, the wind farm still generates, but the solar plants produce nothing and no longer contribute to voltage regulation. The thermal generation at bus 1 would therefore have to be ramped up to meet the increase in load at bus 5. This pattern of load and generation results in large power flows from left to right and causes undervoltages at buses 2, 3, and 5. The MVA flow on line 1–5 $S_{1-5} = \sqrt{P_{1-5}^2 + Q_{1-5}^2} = 88.88$ MVA is greater than $S_{1-5}^{max} = 75$ MVA. As one can observe by comparing the reactive power flows at both ends of the lines, the much larger active power flows cause much larger reactive power losses than in the previous example.

Figure 8.15 illustrates how these operational issues can be resolved. Generating 35 MW with the thermal power plant at bus 2 reduces the power that needs to be provided by the plant at bus 1. The flow on line 1–5 is then $S_{1-5} = 74.37$ MVA $\leq S_{1-5}^{max}$. Injecting 30 Mvar using the reactive power compensation device at bus 5, along with the reactive support provided by the thermal power plant at bus 2, brings all voltage magnitudes back above their lower limit.

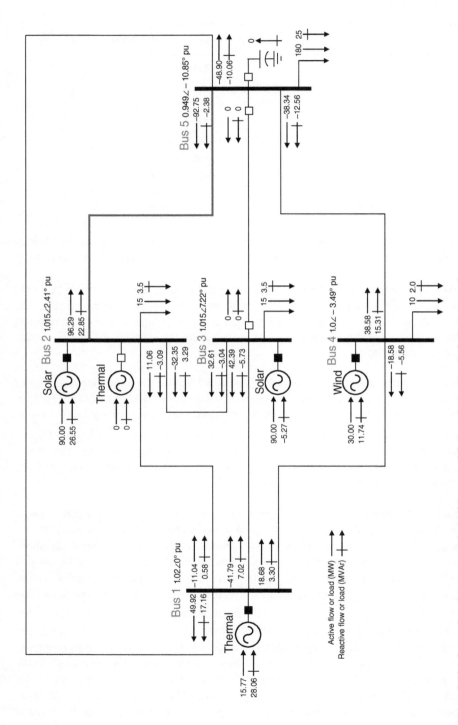

Figure 8.12 Power flow for the conditions of Example 8.8 (i.e., midday, line 3–5 disconnected).

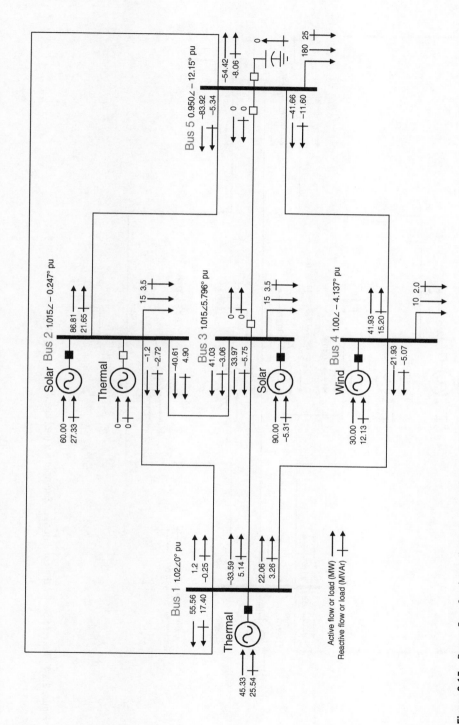

Figure 8.13 Power flow for the redispatched conditions of Example 8.8.

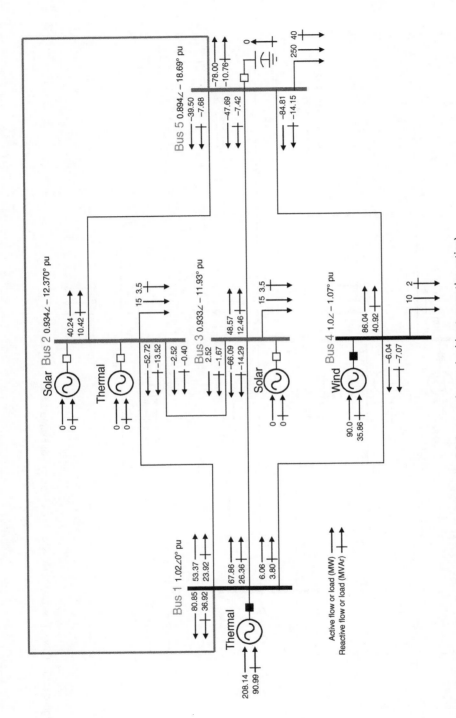

Figure 8.14 Power flow for the conditions of Example 8.9 (i.e., evening peak without corrective action).

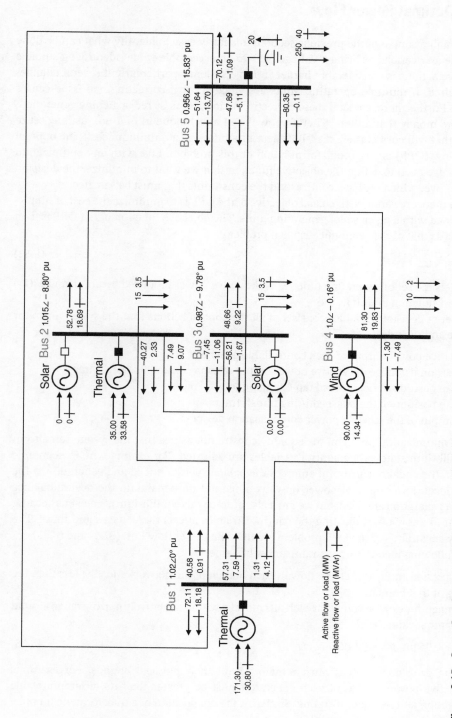

Figure 8.15 Power flow for the conditions of Example 8.9 with corrective actions.

8.8   Optimal Power Flow

In the small example of the previous section, it is fairly easy to identify what can be done to relieve an overload or bring a voltage back to an acceptable value. Identifying suitable corrective actions is considerably harder in large systems, particularly if several changes are required, if multiple operating limits are violated, or if correcting one issue causes another. Furthermore, some of these corrective actions, such as redispatching generation, cost more money than others. Therefore, we need a technique to find not just any set of actions that will correct the issues, but the set that does so at minimum cost. The optimal power flow (OPF) is a rigorous formulation of this problem. Like with any optimization problem, we need to define the objective function that we want to minimize, the decision variables over which we have control, and the constraints that must be satisfied.

While there are variants, the usual objective of the OPF is to minimize the cost of supplying the load with a given set of generating units. The objective function of the OPF is then the same as that of the economic dispatch problem:

$$C = \sum_{i \in G} C_i(P_i) \tag{8.52}$$

where G is the set of generating units, P_i is the active power produced by unit i, and $C_i(P_i)$ is the cost function of unit i.

The set of decision variables consists of all the control actions that the system operator can take and includes:

- The active power output of each controllable generator
- The terminal voltage or reactive power output of each controllable generator
- The tap position of the on-load tap-changing transformers
- The tap position of the phase-shifting transformers
- The output of the reactive power compensation devices.

In an optimization problem, the equality constraint ensures that the system remains in an equilibrium state as the control variables are adjusted. For example, in an economic dispatch, the equality constraint enforces the balance between load and generation at the system level. In an optimal power flow, we must not only maintain the overall balance between generation and load, but we must also enforce this equilibrium at each bus because we want to know how adjusting the control variables affects the voltages and flows. The equality constraints of the OPF problem are thus the power flow Eqs. (8.22) and (8.23).

The following inequality constraints must be respected:

- As expressed in Eq. (8.51), the flow in each branch of the network must be less than the rating of that branch
- The magnitude of the voltage at each bus must be within a relatively narrow range around its nominal value. Typically:

$$0.95 \, \text{pu} \leq V_i \leq 1.05 \, \text{pu} \quad \forall i \tag{8.53}$$

- All the decision variables must remain within their physical bounds. For example, the active power output of each generator must be greater than its minimum stable generation and less than its rating. Similarly, the tap position of a transformer cannot be adjusted beyond what is feasible.

Solving an OPF for actual power systems can be challenging because the number of variables and constraints is large and the relations between them are non-linear. Interior point methods are the state-of-the-art techniques used to solve the full non-linear formulation of the problem described above. A simplification of the problem that is commonly used in electricity markets consists in ignoring the constraints on the voltages and limiting the set of decision variables to the active power output of the generators. The power flow equations can then be linearized, and the problem can be solved using very efficient and robust linear programming techniques.

Further Reading

Gómez-Expósito, A., Conejo, A., and Cañizares, C. (ed.) (2018). *Electric Energy Systems – Analysis and Operation*, 2e. CRC Press.

Grainger, J. and Stevenson, W. (1994). *Power System Analysis*. McGraw Hill.

Power System Analysis and Design, 6, by J. Duncan Glover, Thomas Overbye, Mulukutla S. Sarma, Cengage Learning, 2016.

Problems

P8.1 Build the admittance matrix of the network shown in Figure P8.1. All the values given are impedances.

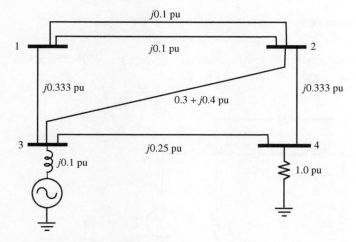

Figure P8.1 Network for problem P8.1.

P8.2 Update the admittance matrix of problem P8.1 to reflect the removal of one of the lines between buses 1 and 2.

P8.3 Sketch the one-line diagram of a network which has an admittance matrix with the structure shown below. ⊗ represents a non-zero element. Don't forget to indicate the bus numbers.

$$Y = \begin{pmatrix} \otimes\otimes\otimes\otimes \\ \otimes\otimes & \otimes \\ \otimes & \otimes & \otimes \\ \otimes & \otimes\otimes \\ \otimes\otimes\otimes\otimes \end{pmatrix}$$

P8.4 Consider the circuit of Figure P8.4. Convert the voltage sources to equivalent current sources and write the nodal equations in matrix form using bus 0 as the reference bus. Do not solve the equations.

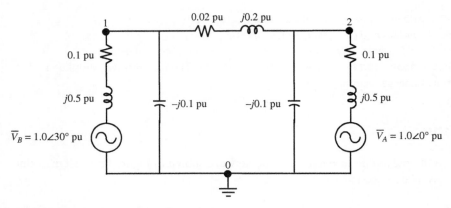

Figure P8.4 Network for problem P8.4.

P8.5 Build the admittance matrix of the network of Figure P8.5. Write the nodal equations for this network.

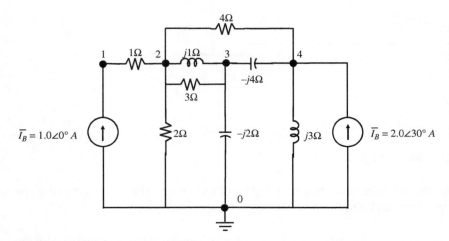

Figure P8.5 Network for problem P8.5.

P8.6 Using a computer, solve the nodal equations that you wrote for problem P8.5.

P8.7 Write the nodal equations for the circuit shown in Figure P8.7.

Figure P8.7 Network for problem P8.7.

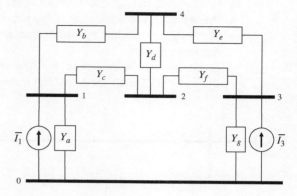

P8.8 Solve the equations that you wrote for problem P8.7 given that:

$$Y_a = -j0.8\ S,\ Y_b = -j4\ S,\ Y_c = -j4\ S,\ Y_d = -j8\ S,\ Y_e = -j5\ S$$

$$Y_f = -j2.5\ S,\ Y_g = -j0.8\ S$$

$$\overline{I_1} = 1.0\angle - 90°A, \overline{I_3} = 0.62\angle - 135°A$$

P8.9 Solve the following equation using the Newton–Raphson method:

$$f(x) = x^3 - x - 1 = 0 \text{ (Start from } x = 1)$$

P8.10 Solve the following equations using the Newton–Raphson method:

$$\begin{cases} f_1(x,y) = x^2 + y^2 - 4 = 0 \\ f_2(x,y) = 1 + x - y = 0 \end{cases} \text{(start from } x = 2, y = 0)$$

P8.11 Use the Newton–Raphson method to solve the following system of non-linear equations:

$$\begin{cases} x^2 - y + 5 = 0 \\ x \cdot y - 50 = 0 \end{cases}$$

Use $x = 2, y = 12$ as the starting point. Show the detail of your first iteration and the convergence record of the first four iterations.

P8.12 Starting from $x = 1$, solve the following equation using the Newton–Raphson method:

$$f(x) = 4 - x^2 - x^3 = 0$$

Do not perform more than three iterations.

P8.13 Consider the small power system shown in Figure P8.13. The series impedance of the line is $Z = j0.25$ pu on a 100 MVA basis. The load connected at bus 2 is 75 MW at 0.95 pf lagging. The generator connected at bus 1 maintains its terminal voltage constant at nominal value and is used as the reference for the angles.

    a. Write the admittance matrix of this network.

    b. Write the power flow equations that need to be solved using the Newton–Raphson method.

    c. Calculate the mismatches for the following condition: $V_2 = 0.95$ pu; $\theta_2 = -12°$; Calculate the Jacobian for these same conditions.

Figure P8.13 Power system for problems P8.13 and P8.14.

P8.14 Consider the small power system shown in Figure P8.13. The series impedance of the line is $Z = j0.2$ pu. on a 100 MVA basis. The load connected at bus 2 is 0.4 pu at 0.8 pf lagging. The generator connected at bus 1 maintains its terminal voltage constant at nominal value and produces the balance of the power.

    (a) Write the admittance matrix of this network.

    (b) Identify all the known and unknown quantities in this system.

    (c) Write the power flow equations for this system. Identify the implicit and explicit equations.

    (d) Write the Jacobian for these equations.

    (e) Perform two iterations of the Newton–Raphson method (by hand) to solve the implicit equations. To do this, you will need to calculate the active and reactive mismatches, calculate the Jacobian matrix, invert this Jacobian matrix, calculate the change in voltage magnitude and angle, and calculate the new values for the voltage magnitude and angle. Use a flat start as your starting point.

    (f) Calculate the active and reactive power produced at the slack bus.

P8.15 Consider the small power system shown in Figure P8.13. On a 100 MVA basis the series impedance of the line is $Z = j0.2$ pu, and the shunt susceptance at each end of the line is $\frac{Y_C}{2} = 0.1$ pu. The load connected at bus 2 is $\overline{S}_2 = 0.5 + j0.15$ pu. The generator connected at bus 1 maintains its terminal voltage constant at nominal value and produces the balance of the power. Calculate the voltage magnitude and angle at bus 2, the power injected into the line at bus 1, and the active and reactive losses in the line. Compare these results with those of problem P8.14.

P8.16 Download one of the freely available power flow programs and use it to replicate the computations of Examples 8.7–8.9. Freely available power flow programs include:

- Matpower: https://matpower.org/ (requires MATLAB™)
- PyPSA: https://pypsa.org/ (requires Python)
- PowerSystems.jl: https://github.com/NREL-SIIP/PowerSystems.jl (requires Julia)
- PowerWorld: https://www.powerworld.com/download-purchase/demo-software (requires Windows™)
- Pandapower: http://www.pandapower.org/ (requires Python)

  The data for these examples can be downloaded in various formats from: https://github.com/Power-Systems-Textbook/TextbookSimulations

P8.17 Using the power flow program that you have downloaded as part of P8.16, run the case corresponding to Figure 8.13. Then modify it to reflect an anticipated wind generation of 150 MW. By how much will this wind generation need to be curtailed if line 4–5 is rated at 95 MVA?

P8.18 Taking as base case the corrected evening peak conditions illustrated in Figure 8.15, use a power flow program to identify the three contingencies that would result in an overload on line 1–3. Which of these contingencies also result in a voltage smaller than 0.95 pu? (Hint: instead of trying all the possible line contingencies, study the base case in Figure 8.15 to identify the contingencies that are most likely to cause this overload).

P8.19 Consider the conditions of Example 8.9 before corrective actions, as illustrated in Figure 8.14. Using a power flow program, study the effect of building an identical line in parallel with line 1–5. Describe the changes in power flows. Would this addition remove all violations of operational limits under the conditions of Example 8.9? If not, what additional action would be required?

P8.20 Consider the conditions of Example 8.7, as illustrated in Figure 8.11. Assuming that line 1–5 is disconnected for maintenance, use a power flow program to determine by how much the active power load at bus 5 can be increased before the flow on one of the lines reaches its limit.

P8.21 Consider the 12-bus power system shown in Figure P8.21. The power flow data for this system is given below on a 100 MVA basis. This data can also be downloaded in formats used by some of the power programs mentioned in P8.16 from: https://github.com/Power-Systems-Textbook/TextbookSimulations

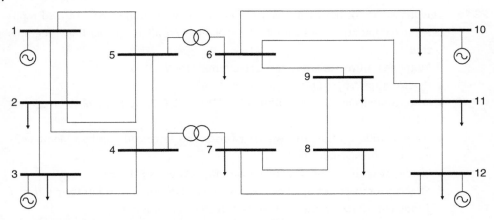

Figure P8.21 Power system for problems P8.21 to P8.28.

Bus data:

Bus	Type	P_{load} (MW)	Q_{load} (MVAr)
1	Slack	25	10
2	PQ	40	15
3	PV	45	16
4	PQ	0	0
5	PQ	0	0
6	PQ	20	7
7	PQ	33	10
8	PQ	20	8
9	PQ	15	6
10	PV	5	2
11	PQ	60	15
12	PV	15	5

Generator data:

Bus	P_{gen} (MW)	V_{gen} (pu)
1	Slack	1.02
3	42	1.02
10	27	1.02
12	33	1.02

Branch data:

From bus	To bus	R	X	B
1	2	0.03876	0.11834	0.0264
1	2	0.03876	0.11834	0.0264
1	5	0.05203	0.20304	0.0492
2	3	0.04699	0.19797	0.0438
2	4	0.05811	0.17632	0.034
2	5	0.05695	0.07388	0.0346
3	4	0.06701	0.17103	0.0128
4	5	0.01335	0.04211	0.0391
4	7	0.0	0.55618	0.0
5	6	0.0	0.55618	0.0
6	9	0.09498	0.1989	0.0176
6	10	0.12291	0.25581	0.0267
6	11	0.06615	0.13027	0.031
7	8	0.03181	0.0845	0.041
7	12	0.12711	0.27038	0.0323
8	9	0.08205	0.19207	0.026
10	11	0.12092	0.19988	0.019
11	12	0.17093	0.34802	0.036

Solve the power flow for these conditions. Since bus 1 is the slack bus, any imbalance between load and the scheduled generation is compensated by generation at this bus.

Analyze the results, i.e., describe the pattern of active and reactive power flows, identify the heavily loaded branches and the buses where the voltage magnitude is significantly below its nominal value. The results of this power flow will provide the base case for subsequent problems.

P8.22 Modify the input data of the 12-bus system of Problem 8.21 to increase the active power generation to 50 MW at bus 10 and at bus 12. Run a power flow for these conditions. Compare the results with the base case to identify the most significant change in the active power flows.

P8.23 Modify the input data of the 12-bus system of Problem 8.21 to increase the reactive power load at bus 11 by 25 Mvar. Run a power flow for these conditions. Compare the results with the base case to identify the most significant changes.

P8.24 Consider again the 12-bus system of Problem 8.21. Run power flow computations to study the effect of adjusting the tap ratio of transformer 4–7 by ±15%. Note that the way a change in tap position is reflected in the input data depends on the power flow program that you use.

P8.25 Consider again the 12-bus system of Problem 8.21. Assume that transformers 4–7 and 5–6 are phase-shifting transformers. Use power flow computations to study the effects of the following adjustments on the active power flows through these transformers. Note that the way a change in phase angle is reflected in the input data depends on the power flow program that you use. Do not change the tap position of these transformers.
- Inserting a phase shift of +15° on transformer 4–7
- Inserting a phase shift of −15° on transformer 4–7
- Inserting a phase shift of +15° on transformer 5–6
- Inserting a phase shift of +15° on both transformers.

P8.26 Consider again the 12-bus system of Problem 8.21. Simulate the outage of transformer 5–6 and note the increase in the active power flow on transformer 4–7. Determine by how much the generation at bus 10 should be adjusted to limit this flow to 100 MW. By how much should the generation at bus 12 be adjusted to achieve the same result? Why are these two amounts slightly different?

P8.27 The outage of transformer 5–6 in Problem P8.26 also creates some violations of the lower voltage limit, even when the active power flow through transformer 4–7 has been reduced to 100 MW. Suggest and quantify a way to correct these violations.

P8.28 Consider again the 12-bus system of Problem 8.21. Determine two ways of reducing the active power flow through transformer 4–7 to 45 MW. Given that the marginal cost of production of generators at buses 1, 10, and 12 are 10 $/MWh, 15 $/MWh, and 25 $/MWh, which of these options would cheaper? Comment briefly on this result.

9

Analysis of Balanced Faults

9.1 Overview

Various types of insulating materials are used to isolate the conducting elements of power equipment from each other and from the ground. For example, each phase of an overhead transmission line must be separated from the other phases and from the earth by a sufficient distance in air. At each tower, these conductors are suspended by strings of insulators made of porcelain, glass, or a polymer. Underground cables are insulated using polyvinyl chloride (PVC), oil-impregnated paper, or cross-linked polyethylene (XLPE). Oil-impregnated paper is also used to provide insulation inside generators and power transformers. All these forms of insulation can and will occasionally fail. Lightning strikes ionize the air, creating an arc, which is a plasma that provides a conductive path. Over time, the accretion of dirt or salt from maritime air reduces the ability of insulator strings to withstand high voltages. Underground cables fail due to either the natural aging of their insulating material or the careless operation of an excavator. Any insulation failure creates an undesirable electrical path through which a potentially very large fault current can flow. These fault currents can cause deaths, injuries, fires, equipment damage, and instabilities in the power system. While the probability of occurrence of faults can be reduced through good design of equipment, their regular maintenance, and a timely replacement as they approach the end of their expected life, it cannot be reduced to zero. It is therefore essential to know what the potential consequences of faults are. Fault currents should indeed be large enough to be distinguishable from normal load currents but small enough to be interruptible through the action of circuit breakers. Fault analysis thus guides the design of power networks, in particular the selection of suitable circuit breakers.

An insulation failure can cause a short circuit between all three phases, between one phase and the ground, between two phases, or between two phases and the ground. Figure 9.1. illustrates these different types of faults. A fault that affects equally all three phases is said to be symmetrical or balanced. Since such a fault does not alter the three-phase symmetry of the system, it can be studied using a single-phase equivalent circuit. The other types of faults do not affect all three phases and thus break this three-phase symmetry. In this chapter, we develop techniques to calculate balanced, three-phase faults. In the next chapter we will introduce symmetrical components, which are a variable transformation that simplifies the analysis of asymmetrical or unbalanced faults.

Power Systems: Fundamental Concepts and the Transition to Sustainability, First Edition. Daniel S. Kirschen.
© 2024 John Wiley & Sons Ltd. Published 2024 by John Wiley & Sons Ltd.
Companion website: www.wiley.com/go/kirschen/powersystems

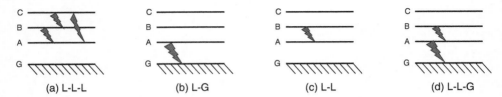

Figure 9.1 Types of faults: (a) balanced three-phase; (b) phase-to-ground; (c) phase-to-phase; (d) two-phase-to-ground.

Depending on the study (Bollen, 1989), single-phase-to-ground faults represent 70–90% and phase-to-phase faults 10–20% of all faults. Two-phase-to-ground and balanced three phase faults account for only 5–15% of the total number. While balanced three-phase faults are rare, they are typically the most severe.

If an insulation failure creates a zero-resistance short circuit, the fault is said to be solid or bolted. On the other hand, if the insulation failure results in an arc or if the path of the short circuit involves some low conductivity material, the fault itself has a resistance R_f. Since the value of this fault resistance is impossible to predict and since neglecting it produces an upper bound of the fault current, we typically assume that faults are solid when calculating fault currents. However, when a fault with a large R_f does occur, the fault current can be small and difficult to detect.

9.2 Two Simple Examples

We will start our study of fault analysis with simple examples and progressively add features to make the results more realistic and the technique applicable to large systems. In particular, we will assume for now that we can model generators as ideal ac voltage sources behind a constant source reactance. We discuss the modeling of generators in more detail in Section 9.4.

Example 9.1 *Fault calculation in a system with one source* Consider the one-line diagram of a simple power system shown in Figure 9.2a. Table 9.1 gives the per unit value of the series reactance of each component on a 50 MVA basis. For simplicity we neglect the resistance and shunt impedances of all the components. Figure 9.2b shows the corresponding circuit diagram. Let us also assume that there are no loads in this system, and that the pre-fault voltages at every point in the system are equal to their nominal values. The magnitude of the internal e.m.f. of the generator is therefore also 1.0 pu.

Let us calculate the fault currents that would result if a balanced three-phase fault were to occur at various locations in this system. If this fault were to be a bolted fault and happen at the terminals of the generator (i.e., at node A), the fault current would be limited only by the internal reactance of the generator:

$$\overline{I}_A^F = \frac{1.0\angle 0°}{j0.5} = 2.0\angle -90° \text{ pu}$$

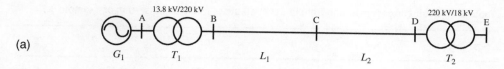

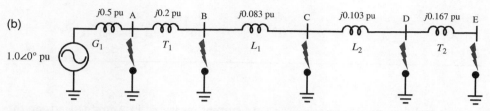

Figure 9.2 (a) One-line diagram for Example 9.1; (b) corresponding circuit diagram showing various potential fault locations.

Table 9.1 Per unit reactance of the components of the system of Example 9.1.

Component	Reactance (pu)
G_1	0.5
T_1	0.2
L_1	0.083
L_2	0.103
T_2	0.167

If, instead of being solid, a fault at bus A itself had a resistance R_f, the fault current would be:

$$\overline{I}_A^F = \frac{1.0\angle 0°}{R_f + j0.5}$$

If a bolted fault were to occur at node C, the fault current would be limited by the internal reactance of the generator, and the series reactances of the step-up transformer and of the line connecting nodes B and C:

$$\overline{I}_C^F = \frac{1.0\angle 0°}{j0.5 + j0.2 + j0.083} = 1.28\angle - 90° \text{ pu}$$

Table 9.2 shows the values of the fault current for bolted faults at every node in this system. The values in kA are calculated by multiplying values in per unit values by the applicable base current:

$$I_B = \frac{S_B}{\sqrt{3}V_B^{LL}}$$

From these results, we observe that, as the location of the fault moves away from the source, the per unit value of the fault current decreases because the total reactance between

Table 9.2 Fault currents resulting from faults at all buses in the small system of Example 9.1.

Bus	A	B	C	D	E
$|I_F|$ (pu)	2.0	1.43	1.28	1.13	0.95
V_B (kV)	13.8	220	220	220	18
I_B (A)	2092	131.2	131.2	131.2	1604
$|I_F|$ (A)	4184	188	168	148	1524
SC (MVA)	100	71.5	64	56.5	47.5

the fault and the source increases. The fault current measured in Amps depends on the nominal voltage at the location of the fault. The last line of this table shows the short circuit level at each bus, which is expressed in MVA and is equal to $\sqrt{3}$ times the pre-fault nominal voltage times the fault current in Amps:

$$SC = \sqrt{3}V_B I_F$$

The short circuit level provides a convenient way of comparing the severity of faults at locations in the system that are at different voltage levels. While the short circuit capacity is expressed in MVA, it does not represent the amount of power dissipated in the fault. While fault currents can be quite large, the voltage across a bolted fault is zero, and so is the power dissipated. If the equivalent impedance between a bus and the sources is small, the fault current and the short circuit capacity at that location are large. If the impedance between a node and the sources is small, load currents cause only a small voltage drop. Good power system design requires balancing these opposing requirements, i.e., having enough impedance to limit fault currents to a manageable level but not so much that the voltages drop excessively under normal conditions.

Example 9.2 *Fault calculation in a system with multiple sources* Calculating the fault current in the previous example was easy because there was only one source and some impedances in series. Figure 9.3 shows the one-line and circuit diagrams of a somewhat larger system with multiple generators. We will again assume that, before a fault occurs, no current flows in this system and that the pre-fault voltages are at their nominal value. The internal voltages of the generators are therefore also equal to 1.0 pu. Suppose that we are interested in calculating the current that would flow into a bolted three-phase fault at bus C. As illustrated in Figure 9.3c, to find the value of this current, we can replace the network by its Thevenin equivalent as seen from bus C. From the definition of the Thevenin equivalent, we recall that the Thevenin voltage is the open circuit voltage, i.e., the pre-fault voltage at bus C. Since we assumed that before the fault the voltage is at its nominal value at all buses, we have $\overline{V_C^{TH}} = 1.0\angle0°$ p.u. The Thevenin impedance is the impedance seen looking into the circuit from bus C when all the voltage sources have been replaced by short circuits.

$$Z_C^{TH} = j(0.33 + 0.167 + 0.103) \parallel j(0.5 + 0.2 + 0.083) \parallel j(0.275 + 0.143) = j0.187 \text{ pu}$$

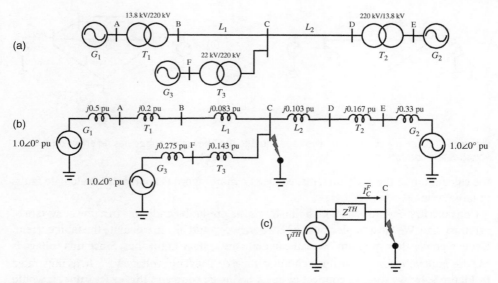

Figure 9.3 (a) One-line diagram for Example 9.2; (b) corresponding circuit diagram; (c) Thevenin equivalent from bus C.

Hence, we have:

$$\overline{I_C^F} = \frac{\overline{V_C^{TH}}}{Z_C^{TH}} = \frac{1.0\angle0°}{j0.187} = 5.348\angle-90° \text{ pu}$$

We can use the same approach to calculate the fault current at other buses. For example, for a fault a bus B, we have:

$$\overline{V_B^{TH}} = 1.0\angle0° \text{ pu}$$

$$Z_B^{TH} = j(0.5+0.2) \parallel \{j0.083 + [j(0.275+0.143) \parallel j(0.33+0.167+0.103)]\} = j0.224 \text{ pu}$$

$$\overline{I_B^F} = \frac{\overline{V_B^{TH}}}{Z_B^{TH}} = \frac{1.0\angle0°}{j0.224} = 4.465\angle-90° \text{ pu}$$

The reader may have noticed that the system illustrating this example is an expanded version of the previous example. Since it contains additional sources and parallel paths from these sources to the location of the fault, it is not surprising that the magnitude of the current for a fault at bus C is significantly higher than the value we calculated in the previous example.

9.3 Balanced Fault Calculations in Large Systems

In the previous examples, we neglected the load currents flowing in the system and assumed that all the voltages, including the internal voltages of the generators, were equal to their nominal value. While these assumptions do not create significant inaccuracies in

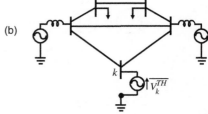

Figure 9.4 Adding a source with a voltage equal to the Thevenin voltage has no effect on the voltages and currents in a network.

the calculation of the fault currents, we need a more rigorous approach to calculate faults in large systems.

Consider first Figure 9.4a, which illustrates the pre-fault conditions in a power system of arbitrary size. We have singled out bus k where we would like to calculate the fault current. Using a power flow program, we can calculate the voltage at this bus. Since this voltage is the pre-fault voltage, it is the open circuit voltage or Thevenin voltage $\overline{V}_k^{TH}$. If, as illustrated in Figure 9.4b, we were to connect at bus k a voltage source of this exact value, it would have no effect whatsoever on the pre-fault currents and voltages anywhere in this system. In particular, no current would flow in or out of this source.

Let us now consider how we could model a bolted fault at bus k. Such a fault clamps the voltage at that bus to zero. As shown in Figure 9.5a, we could model it by inserting a zero-impedance branch between bus k and ground. Instead, we can model this zero voltage by connecting a $+\overline{V}_k^{TH}$ voltage source in series with a $-\overline{V}_k^{TH}$ voltage source between bus k and ground, as illustrated in Figure 9.5b.

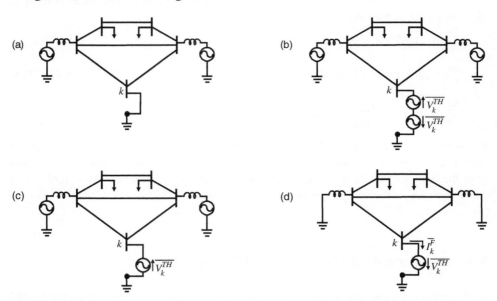

Figure 9.5 Application of the superposition theorem to the calculation of fault currents. (a) Modeling a solid fault as a zero impedance connection; (b) modeling a solid fault as a zero voltage using two equal voltage sources in opposite directions; (c) modeling the pre-fault conditions; (d) modeling for the calculation of the fault current.

As the reader may recall from courses in circuit analysis, the superposition theorem tells us that we can calculate the currents flowing in a linear circuit by adding the currents resulting from different sources acting separately. Applying this theorem to the circuit of Figure 9.5b, we say that the currents flowing in this circuit are equal to the currents flowing in the circuit of Figure 9.5c plus the currents flowing in the circuit of Figure 9.5d. Since Figure 9.5c is identical to Figure 9.4b, the currents in this circuit are the pre-fault currents. In Figure 9.5d, all the voltage sources have been turned off, except the fictitious voltage source $-V_k^{TH}$ connected at the faulted bus. Solving the circuit of Figure 9.5d therefore gives us the current $\overline{I_k^F}$ resulting from a fault at bus k. Since the pre-fault load currents are typically much smaller than the fault currents, they are often neglected in fault calculations.

While the superposition theorem provides a framework for analyzing faults, we need to develop a technique to apply it to the calculation of fault currents in large networks. In Chapter 8, we showed that the network equations could be formulated as follows in matrix form:

$$YV = I \tag{9.1}$$

where Y is the $n \times n$ admittance matrix of the network, V is the vector of nodal voltages, and I is the vector of currents injected into the network at each node. If we apply this equation to the network of Figure 9.5d, we get:

$$Y . \begin{pmatrix} \overline{\Delta V_1} \\ \vdots \\ -\overline{V_k^{TH}} \\ \vdots \\ \overline{\Delta V_n} \end{pmatrix} = \begin{pmatrix} 0 \\ \vdots \\ -\overline{I_k^F} \\ \vdots \\ 0 \end{pmatrix} \tag{9.2}$$

The only current injected in the network is the fault current at node k. It appears here with a negative sign because we have chosen the fault current to be positive when flowing from bus k to ground, which is the opposite of the convention used in nodal analysis. As Figure 9.5d shows, the voltage at node k is known while the voltages at the other nodes are unknown. The $\overline{\Delta V}$ notation is intended to reflect the fact that the fault current creates a voltage drop at these nodes. Equation (9.2) can be rewritten as follows:

$$Z . \begin{pmatrix} 0 \\ \vdots \\ -\overline{I_k^F} \\ \vdots \\ 0 \end{pmatrix} = \begin{pmatrix} \overline{\Delta V_1} \\ \vdots \\ -\overline{V_k^{TH}} \\ \vdots \\ \overline{\Delta V_n} \end{pmatrix} \tag{9.3}$$

where Z is the inverse of Y and is called the impedance matrix of the network. Row k from Equation (9.3) gives:

$$\overline{I_k^F} = \frac{V_k^{TH}}{Z_{kk}} \tag{9.4}$$

where Z_{kk} is element (k, k) of matrix Z. The voltage drops caused by the fault current can then be calculated as follows:

$$\overline{\Delta V_i} = -Z_{ik}\overline{I_k^F} = -\frac{Z_{ik}}{Z_{kk}}\overline{V_k^{TH}} \quad i = 1, \cdots n\ i \neq k \tag{9.5}$$

The voltage at each bus during the fault can then be calculated using the superposition theorem, i.e., by adding these voltage drops to the pre-fault voltages:

$$\overline{V_i^F} = \overline{V_i^{pre}} + \overline{\Delta V_i} = \overline{V_i^{TH}} - \frac{Z_{ik}}{Z_{kk}}\overline{V_k^{TH}} \quad i = 1, \cdots n \tag{9.6}$$

where $\overline{V_i^{pre}}$ is the pre-fault voltage at bus i and $\overline{V_i^F}$ is the voltage at this bus during the fault.

You may have noticed such voltage sags as a momentary dimming of lights when a lightning strike causes a fault on a distribution feeder in your neighborhood. The voltage and the lights return to normal as soon as the fault current is interrupted by the protection system.

Example 9.3 *Fault calculation using the impedance matrix* Figure 9.6 shows the one-line and circuit diagrams of a small power system. For simplicity, we neglect the resistance and shunt susceptance of the branches. We also neglect the loads and assume that all voltages are at nominal value. Therefore:

$$\overline{V_i^{TH}} = 1.0\angle 0° \text{ pu} \quad i = 1, \ldots 4$$

The admittance matrix of this circuit is:

$$Y = \begin{pmatrix} -j26 & j8 & j10 & j5 \\ j8 & -j11 & j3 & 0 \\ j10 & j3 & -j20 & j4 \\ j5 & 0 & j4 & -j9 \end{pmatrix} \text{pu}$$

Inverting this admittance matrix, we get the impedance matrix of the circuit[1]:

$$Z = \begin{pmatrix} j0.182 & j0.174 & j0.151 & j0.168 \\ j0.174 & j0.261 & j0.160 & j0.167 \\ j0.151 & j0.160 & j0.182 & j0.165 \\ j0.168 & j0.167 & j0.165 & j0.278 \end{pmatrix} \text{pu}$$

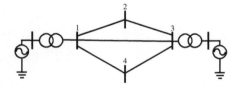
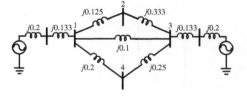

Figure 9.6 One-line and circuit diagrams for Example 9.3. The values indicated are the impedances of the branches in per unit.

---

1 For practical applications, because inverting the admittance matrix of a large power system is computationally expensive, special techniques have been developed to compute only the diagonal elements of the impedance matrix.

For a bolted fault at bus 2, we have:

$$\overline{I}_2^F = \frac{\overline{V}_2^{TH}}{Z_{22}} = \frac{1.0\angle 0°}{j0.261} = 3.83\angle -90° \text{ pu}$$

During the fault the voltages at the other buses are:

$$\overline{V}_1^F = 1.0\angle 0° - \frac{j0.174}{j0.261}1.0\angle 0° = 0.334\angle 0° \text{ pu}$$

$$\overline{V}_3^F = 1.0\angle 0° - \frac{j0.160}{j0.261}1.0\angle 0° = 0.388\angle 0° \text{ pu}$$

$$\overline{V}_4^F = 1.0\angle 0° - \frac{j0.167}{j0.261}1.0\angle 0° = 0.358\angle 0° \text{ pu}$$

9.4 Modeling Generators for Fault Calculations

In Chapter 4 we developed a simple model of the steady-state behavior of synchronous generator. This model is not suitable for fault calculations because it does not account for what happens when the stator currents change rapidly. To develop a better model, we need to have a closer look at how the magnetic flux links the windings during steady-state and transient conditions. In addition to the stator and field windings that we discussed in Chapter 4, Figure 9.7 shows damper windings in the pole faces of the rotor. These windings are in short circuit and, as their name suggest, serve to dampen mechanical oscillations in the position of the rotor. The stator, field, and damper windings are magnetically coupled, which means that if a current flows in one of these windings, it affects the magnetic flux through the other windings. In particular, we saw in Chapter 4 how a current in the field winding creates a field flux whose rotation induces voltages and currents in the stator windings. In turn, the three-phase stator currents create a stator magnetic flux wave that rotates synchronously with the rotor and links the field and damper windings. This stator flux is in the opposite direction of the flux created by the field winding. Figure 9.8a illustrates the path of this

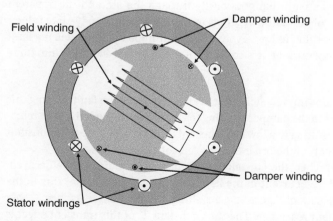

Figure 9.7 Schematic representation of the windings of a synchronous generator.

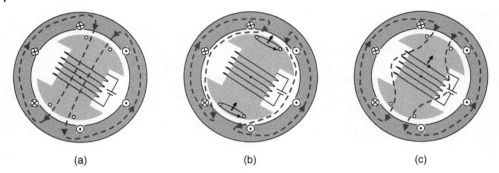

(a)   (b)   (c)

Figure 9.8 Path of the stator flux in: (a) the steady state, (b) the subtransient state, (c) the transient state.

stator flux when the generator is in the steady state. Since this path is mainly through iron in the stator and rotor, it has a small reluctance. The self-inductance of the stator windings L_S is thus large in the steady state, and so is the synchronous reactance $X_S = \omega L_S$.

Let us examine what happens when the stator current suddenly increases as a result of a fault in the network. Since the stator flux opposes the field flux, an increase in the stator current would reduce the net flux through the damper and field winding. However, Lenz's law tells us that a change in flux induces a voltage in these windings and drives a current that opposes the change in flux. Immediately after a fault, the currents induced in the damper windings thus prevent the change in stator flux from entering the rotor, as illustrated in Figure 9.8b. Since the path that this change in stator flux is forced to flow is through the airgap, its reluctance is much higher, and the apparent self-inductance and reactance of the stator windings is thus much lower. Because the currents induced in the damper windings decrease according to the L/R time constant of these windings, their shielding effect decreases, and the stator flux is able to progressively re-enter the rotor, as shown in Figure 9.8c. However, the current induced in the field winding by the change in stator flux decreases more slowly than the currents in the damper windings. Until this current decays, the stator flux follows a path that goes partly through the air, and the apparent self-inductance and reactance of the stator windings remains below its steady-state value.

While the path that the stator flux follows evolves in a continuous manner when a fault causes a sudden change in stator current, it is convenient to divide this evolution into three distinct periods:

- The subtransient period begins immediately after the occurrence of the fault. During this period, the induced currents in the damper winding force the flux through the high reluctance path of Figure 9.8b, resulting in a low subtransient reactance X''. The time constant T'' of this subtransient effect is of the order of a few tens of milliseconds.
- The transient period starts when the currents in the damper windings have decayed. During this period, the induced current in the field winding keeps the stator flux in the somewhat lower reluctance path of Figure 9.8c, resulting in a transient reactance X' that is larger than the subtransient reactance. The time constant T' of this transient effect is of the order of a few hundreds of milliseconds.

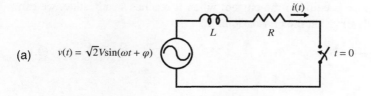

(a) $v(t) = \sqrt{2}V\sin(\omega t + \varphi)$

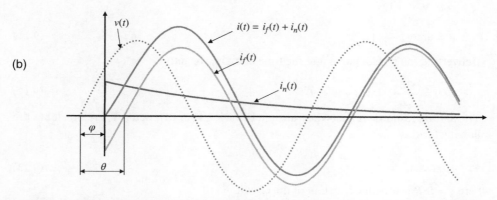

(b)

Figure 9.9 Transient currents following the sudden closure of an RL circuit supplied by a voltage source. $\varphi = 40°$, $\theta = 70°$, and $\tau = 2/3$ T.

- The steady-state period when all the induced currents have decayed and the stator flux flows through the low reluctance path of Figure 9.8a, resulting in the steady-state or synchronous reactance X_S.

Therefore, we have the following relation between the subtransient, transient and synchronous reactances:

$$X'' < X' < X_S \qquad (9.7)$$

Since the subtransient reactance is the smallest, the fault currents will be highest during the subtransient period. For fault calculations, we will therefore model generators as an ideal voltage source behind the subtransient reactance.

So far, we have assumed that the fault currents are sinusoidal. However, fault currents also contain a transient dc component. To understand how this component arises, consider the simple RL circuit shown in Figure 9.9a. Closing the switch at time $t = 0$ is similar to applying a fault at the terminals of a generator. If we apply KVL to this circuit after the switch is closed, we get:

$$L\frac{di(t)}{dt} + Ri(t) = \sqrt{2}V \sin (\omega t + \varphi) \quad t \geq 0 \qquad (9.8)$$

From circuit analysis, we recall that the solution of this differential equation is the sum of a forced response and a natural response:

$$i(t) = i_f(t) + i_n(t) \qquad (9.9)$$

Since the forced response is equal to the current when it reaches steady state, we can calculate it by solving this circuit in the phasor domain:

$$\bar{I} = \frac{\bar{V}}{R + j\omega L} = \frac{V\angle\varphi}{Z\angle\theta} = \frac{V}{Z}\angle(\varphi - \theta) \tag{9.10}$$

where:

$$Z = \sqrt{R^2 + (\omega L)^2}$$

$$\theta = \text{arctg}\left(\frac{\omega L}{R}\right) \tag{9.11}$$

Converting the phasor current to the time domain, we get:

$$i_f(t) = \frac{\sqrt{2}V}{Z}\sin(\omega t + \varphi - \theta) \tag{9.12}$$

The natural response reflects the transient behavior of an RL circuit. Therefore, it has the following form:

$$i_n(t) = Ae^{-\frac{t}{\tau}} \tag{9.13}$$

where $\tau = L/R$ is the time constant of the circuit.

Inserting the expressions (9.12) and (9.13) in (9.9), we get:

$$i(t) = \frac{\sqrt{2}V}{Z}\sin(\omega t + \varphi - \theta) + Ae^{-\frac{t}{\tau}} \quad t \geq 0 \tag{9.14}$$

To determine the unknown constant A, we recall that the current in an inductor cannot change instantaneously. Since this current was zero before the switch closes at $t = 0$, we have:

$$i(0) = \frac{\sqrt{2}V}{Z}\sin(\varphi - \theta) + A = 0 \tag{9.15}$$

Hence:

$$i(t) = \frac{\sqrt{2}V}{Z}\left[\sin(\omega t + \varphi - \theta) - \sin(\varphi - \theta)e^{-\frac{t}{\tau}}\right] \quad t \geq 0 \tag{9.16}$$

Figure 9.9b shows the evolution of this current and its components for particular values of φ, θ, and τ. The natural response $i_n(t)$ introduces a unidirectional component or dc offset whose magnitude depends on when the fault occurs with respect to the voltage waveform. It is maximum for $\varphi - \theta = \pm 90°$ and zero for $\varphi - \theta = 0°$. In many cases, it will thus increase the maximum instantaneous value of the fault current. Since faults occur at random times, we should assume the worst-case scenario when calculating the magnitude of the fault current that must be interrupted. Note that in a three-phase system, the angle φ of the three phases differs by 120°. The magnitude of the dc offset will thus be different in each phase.

Figures 9.10–9.12 illustrate the modeling of generators by showing the waveforms of the current in all three phases if a bolted fault is applied at $t = 0$ to a generator that was

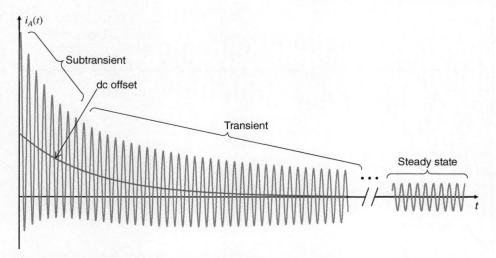

Figure 9.10 Currents in phase A of a synchronous generator following a bolted fault on the terminals of the generator at $t = 0$.

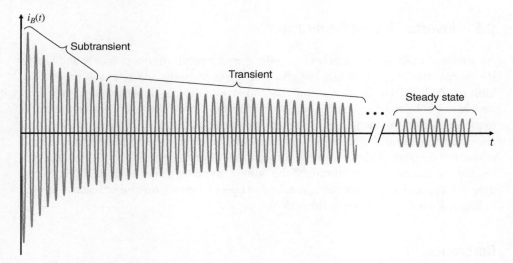

Figure 9.11 Currents in phase B of a synchronous generator following a bolted fault on the terminals of the generator at $t = 0$.

previously operating in open circuit. The magnitude of the ac component of these waveforms exhibits an initial rapid decrease corresponding to the subtransient period, followed by a slower decrease over the transient period until it reaches steady state. The phase angle φ of the voltage waveform has been chosen arbitrarily to create a positive dc offset in phase A, a zero dc offset in phase B, and a negative dc offset in phase C.

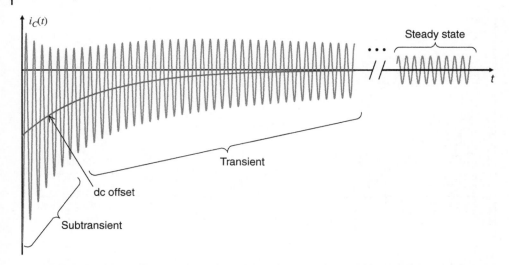

Figure 9.12 Currents in phase C of a synchronous generator following a bolted fault on the terminals of the generator at $t = 0$.

9.5 Inverter-Based Generation

As discussed in Chapter 5, power electronics converters can control precisely and rapidly the current that they inject into the grid. This gives them the ability to self-protect, i.e., limit the fault current as soon as their control system detects an abnormal voltage drop, such as might be caused by a fault. While limiting the current from the inverter is essential to prevent a catastrophic failure of the semiconductor switches, fully disconnecting these resources from the grid is undesirable because it might create a power imbalance and exacerbate the problem. Rules governing the connection of these resources to the grid thus require the inverter to "ride through" the fault if the voltage drop is mild. If this voltage drop is more severe, the inverter is required to inject a limited amount of reactive current to help maintain the stability of the system.

Reference

Bollen, M.H.J. (1989). *On Travelling-Wave-Based Protection of High-Voltage Networks.* Technische Universiteit Eindhoven https://doi.org/10.6100/IR316599.

Further Reading

Grainger, J. (1994). *Power System Analysis.* William Stevenson: McGraw Hill.
Glover, J.D., Overbye, T., and Sarma, M.S. (2016). *Power System Analysis and Design,* 6e. Cengage Learning.
Tleis, N. (2019). *Power Systems Modelling and Fault Analysis: Theory and Practice,* 2e. Academic Press.

Problems

P9.1 Figure P9.1 shows the one-line diagram of a distribution feeder. This feeder is supplied from a transmission network that maintains the voltage on the 115 kV side of T_1 at nominal value. The table below gives the parameters of the various components.

(a) Convert the parameters to consistent per unit values assuming a 10 MVA basis.

(b) Draw a circuit diagram for this network.

(c) Assuming that there are no loads on this system, calculate the fault currents in per unit and in Amps at buses B, C, D, E, and F.

(d) Discuss your results.

Component	Rating	Reactance
T_1	5 MVA	0.2 pu
T_2	2 MVA	0.15 pu
T_3	50 kVA	0.1 pu
L_1	—	2 Ω
L_2	—	3 Ω

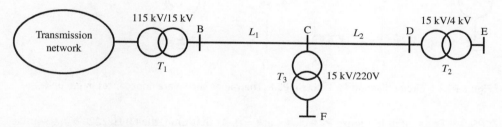

Figure P9.1 Distribution feeder of Problem P9.1.

P9.2 Calculate the per unit fault current for a bolted, balanced three-phase fault at bus A of Figure 9.3. Given that the base power is 50 MVA, convert this value to Amps. Calculate the short circuit level at that bus. Neglect loads and assume that the voltages are at nominal value pre-fault.

P9.3 Repeat problem P9.1 for a bolted, balanced three-phase fault at bus D.

P9.4 Repeat problem P9.1 for a bolted, balanced three-phase fault at bus E.

P9.5 Repeat problem P9.1 for a bolted, balanced three-phase fault at bus F.

P9.6 (a) Build the admittance matrix of the system of Problem P9.1.

(b) Invert this matrix to obtain the impedance matrix of this system.

(c) Using the impedance matrix, calculate the fault currents for faults at buses B, C, D, E, and F

(d) Compare these results with the results of Problem P9.1.

P9.7 Using the impedance matrix calculated in Problem P9.6, calculate the voltage that would result at buses B, C, E, and F during balanced bolted faults at buses D and F. Briefly discuss your results.

P9.8 Consider the small system shown in Figure P9.8. Assume that there are no loads and that all voltages are at nominal value.
(a) Calculate the fault currents in per unit for balanced bolted faults at buses A and D using impedance matrix of this network.
(b) Convert these fault currents to SI units assuming that the base power is 20 MVA and the base voltage is 30 kV.
(c) Calculate the voltage that would result at buses B and C before these faults clear.
(d) Briefly discuss your results.

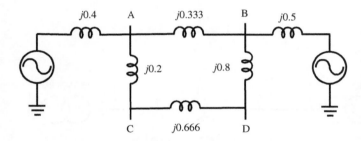

Figure P9.8 Circuit diagram for Problem P9.8. The values shown are impedances in per unit.

P9.9 The switch in Figure P9.9 closes at $t = 0$. At that time, the 60 Hz, 220 V_{RMS} source has a phase of 30°. Given $L = 2\,H$ and $R = 50\,\Omega$, calculate the instantaneous value of the current $i(t)$ at $t = 0.05$ s.

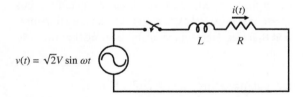

Figure P9.9 Circuit diagram for Problem P9.9.

P9.10 Using the same data as in Problem P9.9 calculate the instantaneous current at $t_2 = 50$ ms assuming that the switch closes at $t_1 = 37.5$ ms.

P9.11 A three-phase, 300 MVA, 30 kV, 60 Hz synchronous generators has the following parameters:

$X_d'' = 0.2$ pu	$T_d'' = 0.04$ s
$X_d' = 0.4$ pu	$T_d' = 2.5$ s
$X_d = 1.5$ pu	$T_A = 0.3$ s

(All per unit reactances are given on the basis of the rating of the generator).
Assume that this generator is unloaded and operates at nominal voltage.

(a) Calculate the fault current in per unit and in kA immediately after the fault.

(b) Write an expression showing how the worst-case dc offset current evolves over time.

(c) Write an expression showing how the ac fault current evolves over time.

(d) What would be the RMS value of the ac component of the fault current that a circuit breaker would have to interrupt to clear the fault in four cycles?

(e) What would be the maximum dc offset of the fault current at that time?

(f) Calculate the rms value of the combined ac and dc components of this fault current at that time.

10

Analysis of Unbalanced Faults

10.1 Overview

As we mentioned in the introduction to Chapter 9, the vast majority of faults in power systems do not affect equally all three phases. Since the occurrence of such faults creates an imbalance in the system, they cannot be analyzed using a single-phase representation of the system. Studying unbalanced faults using the phase variables is hard and does not provide much physical insight. To simplify this analysis, we will instead transform the phase variables into new variables called symmetrical components. This chapter thus begins by explaining how to transform phase voltages and currents into their symmetrical components. It then discusses how using these symmetrical components decouples a three-phase network into three separate networks, each of which representing a symmetrical component. Building on this foundation, it then shows how to analyze different types of unbalanced faults.

10.2 Symmetrical Components

10.2.1 Notation

Explaining symmetrical component is simpler if we introduce the shorthand notation a, which is defined as follows:

$$a = 1\angle 120° = -0.5 + j0.866 \tag{10.1}$$

Multiplying a phasor by a thus increases its phase by 120° while leaving its magnitude unchanged. Applying the operator a twice means multiplying the phasor by a^2, i.e., shifting its phase by 240°.

$$a^2 = 1\angle 240° = -0.5 - j0.866 \tag{10.2}$$

Applying the operator a three times leaves a phasor unaffected because:

$$a^3 = 1\angle 360° = 1\angle 0° = 1 \tag{10.3}$$

From (10.1) and (10.2) we get the following important relation:

$$1 + a + a^2 = 0 \tag{10.4}$$

Power Systems: Fundamental Concepts and the Transition to Sustainability, First Edition. Daniel S. Kirschen.
© 2024 John Wiley & Sons Ltd. Published 2024 by John Wiley & Sons Ltd.
Companion website: www.wiley.com/go/kirschen/powersystems

10.2.2 Concept of Symmetrical Components

C.L. Fortescue (1918) demonstrated that any unbalanced three-phase voltage or current can be decomposed into the sum of three balanced three-phase voltages or currents called sequence components:

- A zero-sequence component where all three voltages or currents have the same phase
- A positive-sequence component where the three voltages or currents are 120° out of phase with each other in the positive phase sequence (e.g., a–b–c)
- A negative-sequence component where the three voltages or currents are 120° out of phase with each other but in the opposite phase sequence (e.g., a–c–b).

We will use the subscripts "0," to denote the zero-sequence, "1" for the positive-sequence, and "2" for the negative-sequence. Figure 10.1 illustrates these sequence components using voltage phasors.

Each of these components is balanced in the sense that what happens in each phase is repeated in the other two phases with either a 120° phase difference (for the positive- and negative-sequence components) or a 0° phase difference (for the zero-sequence component). This observation allows us to analyze each component as if it consisted of only one phase, although with a different set of impedances for each sequence component, and to represent each sequence component using a single phasor:

For the zero-sequence component:

$$\overline{V_0} = \overline{V_{a,0}} = \overline{V_{b,0}} = \overline{V_{c,0}} \tag{10.5}$$

For the positive-sequence component:

$$\overline{V_1} = \overline{V_{a,1}}; \quad \overline{V_{b,1}} = a^2\overline{V_1}; \quad \overline{V_{c,1}} = a\overline{V_1} \tag{10.6}$$

For the negative-sequence component:

$$\overline{V_2} = \overline{V_{a,2}}; \quad \overline{V_{b,2}} = a\overline{V_2}; \quad \overline{V_{c,2}} = a^2\overline{V_2} \tag{10.7}$$

10.2.3 Calculating the Sequence Components

Stating that an unbalanced three-phase voltage can be decomposed into the sum of its zero-, positive-, and negative-sequence components means that each of the phase voltages is equal

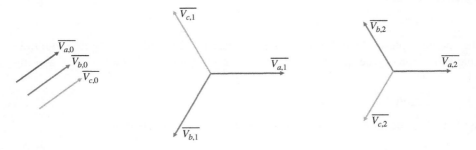

Zero-sequence   Positive-sequence   Negative-sequence

Figure 10.1 Phasor representation of sequence components.

to the sum of the corresponding phase voltage of each component:

$$\overline{V_a} = \overline{V_{a,0}} + \overline{V_{a,1}} + \overline{V_{a,2}}$$
$$\overline{V_b} = \overline{V_{b,0}} + \overline{V_{b,1}} + \overline{V_{b,2}}$$
$$\overline{V_c} = \overline{V_{c,0}} + \overline{V_{c,1}} + \overline{V_{c,2}} \tag{10.8}$$

Inserting (10.5)–(10.7) into (10.8), we get:

$$\overline{V_a} = \overline{V_0} + \overline{V_1} + \overline{V_2}$$
$$\overline{V_b} = \overline{V_0} + a^2\overline{V_1} + a\overline{V_2}$$
$$\overline{V_c} = \overline{V_0} + a\overline{V_1} + a^2\overline{V_2} \tag{10.9}$$

We can put (10.9) in matrix form:

$$\begin{pmatrix} \overline{V_a} \\ \overline{V_b} \\ \overline{V_c} \end{pmatrix} = \begin{pmatrix} 1 & 1 & 1 \\ 1 & a^2 & a \\ 1 & a & a^2 \end{pmatrix} \begin{pmatrix} \overline{V_0} \\ \overline{V_1} \\ \overline{V_2} \end{pmatrix} \tag{10.10}$$

If we set:

$$\boldsymbol{A} = \begin{pmatrix} 1 & 1 & 1 \\ 1 & a^2 & a \\ 1 & a & a^2 \end{pmatrix} \tag{10.11}$$

The inverse of matrix $\boldsymbol{A}$ is:

$$\boldsymbol{A}^{-1} = \frac{1}{3}\begin{pmatrix} 1 & 1 & 1 \\ 1 & a & a^2 \\ 1 & a^2 & a \end{pmatrix} \tag{10.12}$$

Hence, we can rewrite (10.10) as follows:

$$\begin{pmatrix} \overline{V_0} \\ \overline{V_1} \\ \overline{V_2} \end{pmatrix} = \frac{1}{3}\begin{pmatrix} 1 & 1 & 1 \\ 1 & a & a^2 \\ 1 & a^2 & a \end{pmatrix} \begin{pmatrix} \overline{V_a} \\ \overline{V_b} \\ \overline{V_c} \end{pmatrix} \tag{10.13}$$

which shows how we can calculate the sequence components given the phase voltages. This equation can be expressed in matrix form:

$$\boldsymbol{V_S} = \boldsymbol{A}^{-1}\boldsymbol{V_P} \tag{10.14}$$

where the subscripts $\boldsymbol{S}$ and $\boldsymbol{P}$ denote, respectively, sequence and phase variables.

The transformation from phase variables to symmetrical components and its inverse are also applicable to three-phase currents:

$$\begin{pmatrix} \overline{I_a} \\ \overline{I_b} \\ \overline{I_c} \end{pmatrix} = \begin{pmatrix} 1 & 1 & 1 \\ 1 & a^2 & a \\ 1 & a & a^2 \end{pmatrix} \begin{pmatrix} \overline{I_0} \\ \overline{I_1} \\ \overline{I_2} \end{pmatrix} \tag{10.15}$$

And:

$$\begin{pmatrix} \overline{I_0} \\ \overline{I_1} \\ \overline{I_2} \end{pmatrix} = \frac{1}{3}\begin{pmatrix} 1 & 1 & 1 \\ 1 & a & a^2 \\ 1 & a^2 & a \end{pmatrix}\begin{pmatrix} \overline{I_a} \\ \overline{I_b} \\ \overline{I_c} \end{pmatrix}$$

(10.16)

We can write these last two equations compactly as follows:

$$I_P = AI_S$$

(10.17)

And:

$$I_S = A^{-1}I_P$$

(10.18)

Example 10.1 *Sequence Components of a Balanced Set of Voltages in the a–b–c Phase Sequence* Calculate the sequence components of the following set of balanced voltages:

$$\overline{V_a} = 110\angle 0° \text{ V}; \quad \overline{V_b} = 110\angle 240° \text{ V}; \quad \overline{V_c} = 110\angle 120° \text{ V}$$

$$\overline{V_0} = \frac{1}{3}(110\angle 0° + 110\angle 240° + 110\angle 120°) = 0$$

$$\overline{V_1} = \frac{1}{3}(110\angle 0° + 1\angle 120° \times 110\angle 240° + 1\angle 240° \times 110\angle 120°)$$

$$= \frac{1}{3}(110\angle 0° + 110\angle 0° + 110\angle 0°)$$

$$= 110\angle 0° \text{ V} = \overline{V_a}$$

$$\overline{V_2} = \frac{1}{3}(110\angle 0° + 1\angle 240° \times 110\angle 240° + 1\angle 120° \times 110\angle 120°)$$

$$= \frac{1}{3}(110\angle 0° + 110\angle 120° + 110\angle 240°) = 0$$

which confirms that we only need the positive-sequence component to represent a balanced set of voltages in the positive phase sequence. Conversely, the following example shows that only the negative-sequence component is needed to represent a balanced set of currents or voltages is in the opposite phase sequence.

Example 10.2 *Sequence Components of a Balanced Set of Currents in the a–c–b Phase Sequence* Calculate the sequence components of the following set of balanced currents:

$$\overline{I_a} = 30\angle 0° \text{ A}; \quad \overline{I_b} = 30\angle 120° \text{ A}; \quad \overline{I_c} = 30\angle 240° \text{ A}$$

The sequence components are:

$$\overline{I_0} = \frac{1}{3}(30\angle 0° + 30\angle 120° + 30\angle 240°) = 0$$

$$\overline{I_1} = \frac{1}{3}(30\angle 0° + 1\angle 120° \times 30\angle 120° + 1\angle 240° \times 30\angle 240°)$$

$$= \frac{1}{3}(30\angle 0° + 30\angle 240° + 30\angle 120°) = 0$$

$$\overline{I_2} = \frac{1}{3}(30\angle 0° + 1\angle 240° \times 30\angle 120° + 1\angle 120° \times 30\angle 240°)$$

$$= \frac{1}{3}(30\angle 0° + 30\angle 0° + 30\angle 0°)$$

$$= 30\angle 0° \ \text{A} = \overline{I_a}$$

In this case, only the negative-sequence component is non-zero.

Example 10.3 *Sequence Components of a Set of Unbalanced Voltages* Calculate the sequence components of the following set of unbalanced voltages:

$$\overline{V_a} = 100\angle 0° \ \text{V}; \quad \overline{V_b} = 90\angle 240° \ \text{V}; \quad \overline{V_c} = 110\angle 120° \ \text{V}$$

$$\overline{V_0} = \frac{1}{3}(100\angle 0° + 90\angle 240° + 110\angle 120°) = 5.77\angle 90° \ \text{V}$$

$$\overline{V_1} = \frac{1}{3}(100\angle 0° + 1\angle 120° \times 90\angle 240° + 1\angle 240° \times 110\angle 120°) = 100\angle 0° \ \text{V}$$

$$\overline{V_2} = \frac{1}{3}(100\angle 0° + 1\angle 240° \times 90\angle 240° + 1\angle 120° \times 110\angle 120°) = 5.77\angle -90° \ \text{V}$$

Since the imbalance between the three phase voltages is relatively small, the positive-sequence component is much larger than the zero- and negative-sequence components.

Example 10.4 *Calculating Phase Voltages from Their Sequence Components* Calculate the phase voltages from the following sequence components:

$$\overline{V_0} = 5\angle 10° \ \text{V}; \quad \overline{V_1} = 50\angle 5° \ \text{V}; \quad \overline{V_2} = 10\angle 30° \ \text{V}$$

$$\overline{V_a} = 5\angle 10° + 50\angle 5° + 10\angle 30° = 64.21\angle 9.16° \ \text{V}$$

$$\overline{V_b} = 5\angle 10° + 1\angle 240° \times 50\angle 5° + 1\angle 120° \times 10\angle 30° = 46.63\angle -122.23° \ \text{V}$$

$$\overline{V_c} = 5\angle 10° + 1\angle 120° \times 50\angle 5° + 1\angle 240° \times 10\angle 30° = 39.71\angle 126.74° \ \text{V}$$

10.2.4 Relation Between the Neutral and Zero-sequence Currents

Consider the Y-connected load shown in Figure 10.2, where the neutral point n is connected to ground. Applying KCL at n, we get:

$$\overline{I_n} = \overline{I_a} + \overline{I_b} + \overline{I_c} \tag{10.19}$$

The first row of matrix Eq. (10.15) is:

$$\overline{I_0} = \frac{1}{3}(\overline{I_a} + \overline{I_b} + \overline{I_c}) \tag{10.20}$$

Combining these last two equations gives:

$$\overline{I_0} = \frac{1}{3}\overline{I_n} \tag{10.21}$$

From which we conclude that a zero-sequence current can occur only if the neutral current is not zero, which requires the existence of a connection between the neutral point and

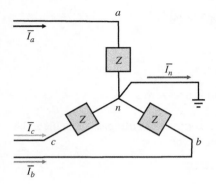

Figure 10.2 Y-connected load with solidly grounded neutral connection.

ground. Conversely, if the currents are balanced, the zero-sequence and neutral currents are both equal to zero.

Example 10.5 *Zero-sequence and Neutral Currents* Given $\overline{I}_a = 60\angle 0°$ A; $\overline{I}_b = 60\angle 240°$ A; $\overline{I}_c = 0$ A, calculate the zero-sequence and neutral currents.

$$\overline{I}_0 = \frac{1}{3}(60\angle 0° + 60\angle 240° + 0) = 20\angle -60°\ \text{A}$$

$$\overline{I}_n = 3\overline{I}_0 = 60\angle -60°\ \text{A}$$

10.3 Sequence Networks

In this section, we apply symmetrical components to the analysis of three-phase circuits. Using the example of loads with balanced impedances, we first show that converting from phase variables to symmetrical components transforms a balanced set of three-phase impedances into three decoupled sequence impedances, one for each of the positive-, negative-, and zero-sequence component. We then discuss how to represent lines, generators, and transformers in terms of sequence impedances. Finally, we show how to combine these sequence impedances into the sequence networks that we will use to analyze unbalanced faults.

10.3.1 Sequence Networks Representation of Impedance Loads

Figure 10.3 shows a balanced Y-connected load where the neutral is connected to ground through an impedance.

Writing KVL around the loop a–n–g, we have:

$$\overline{V}_a = \overline{I}_a Z_Y + \overline{I}_n Z_n \tag{10.22}$$

By applying KCL at node n we can express $\overline{I}_n$ as a function of the phase currents:

$$\overline{I}_n = \overline{I}_a + \overline{I}_b + \overline{I}_c \tag{10.23}$$

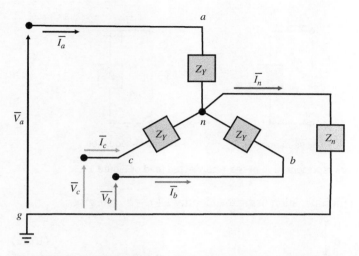

Figure 10.3 Y-connected load with neutral connected to ground through an impedance.

Combining (10.19) and (10.20) gives:

$$\overline{V}_a = \overline{I}_a(Z_Y + Z_n) + \overline{I}_b Z_n + \overline{I}_c Z_n \tag{10.24}$$

Similarly:

$$\overline{V}_b = \overline{I}_a Z_n + \overline{I}_b(Z_Y + Z_n) + \overline{I}_c Z_n \tag{10.25}$$

$$\overline{V}_c = \overline{I}_a Z_n + \overline{I}_b Z_n + \overline{I}_c(Z_Y + Z_n) \tag{10.26}$$

Putting these last three equations in matrix form, we get:

$$\begin{pmatrix} \overline{V}_a \\ \overline{V}_b \\ \overline{V}_c \end{pmatrix} = \begin{pmatrix} Z_Y + Z_n & Z_Y & Z_Y \\ Z_Y & Z_Y + Z_n & Z_Y \\ Z_Y & Z_Y & Z_Y + Z_n \end{pmatrix} \begin{pmatrix} \overline{I}_a \\ \overline{I}_b \\ \overline{I}_c \end{pmatrix} \tag{10.27}$$

which we can write in compact form using the notations defined in (10.14) and (10.17):

$$V_P = Z_P I_P \tag{10.28}$$

Multiplying on the left both sides of (10.28) by the A^{-1} matrix defined in (10.12), we have:

$$A^{-1} V_P = A^{-1} Z_P I_P \tag{10.29}$$

Inserting (10.14) and (10.17) in (10.29), we get:

$$V_S = Z_S I_S \tag{10.30}$$

where:

$$Z_S = A^{-1} Z_P A = \frac{1}{3} \begin{pmatrix} 1 & 1 & 1 \\ 1 & a & a^2 \\ 1 & a^2 & a \end{pmatrix} \begin{pmatrix} Z_Y + Z_n & Z_n & Z_n \\ Z_n & Z_Y + Z_n & Z_n \\ Z_n & Z_n & Z_Y + Z_n \end{pmatrix} \begin{pmatrix} 1 & 1 & 1 \\ 1 & a^2 & a \\ 1 & a & a^2 \end{pmatrix} \tag{10.31}$$

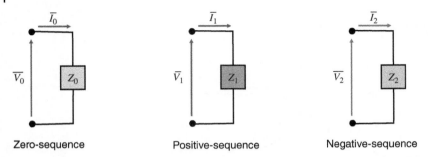

Zero-sequence   Positive-sequence   Negative-sequence

Figure 10.4 Sequence network representation of the Y-connected load of Figure 10.3.

Carrying out this matrix multiplication using the identity $1 + a + a^2 = 0$, gives:

$$\mathbf{Z_S} = \begin{pmatrix} Z_Y + 3Z_n & 0 & 0 \\ 0 & Z_Y & 0 \\ 0 & 0 & Z_Y \end{pmatrix} \tag{10.32}$$

Expanding (10.30) using (10.32), we get:

$$\overline{V}_0 = (Z_Y + 3Z_n)\overline{I}_0 = Z_0\overline{I}_0 \tag{10.33}$$

$$\overline{V}_1 = Z_Y\overline{I}_1 = Z_1\overline{I}_1 \tag{10.34}$$

$$\overline{V}_2 = Z_Y\overline{I}_2 = Z_2\overline{I}_2 \tag{10.35}$$

Figure 10.4 summarizes these equations into three equivalent circuits and illustrates several important conclusions:

- Converting phase quantities into their symmetrical components transforms the coupled Eq. (10.27) into the three decoupled Eqs. (10.33–10.35).
- These equations show that positive-sequence currents create only positive-sequence voltages, negative-sequence currents create only negative-sequence voltages, and zero-sequence currents create only zero-sequence voltages.
- The impedance to positive-sequence currents Z_1 (or positive-sequence impedance for short) is equal to the phase impedance Z_Y.
- The negative-sequence impedance Z_2 is also equal to Z_Y because a passive balanced three-phase impedance load is indifferent to the phase sequence of the applied voltages.
- The neutral impedance Z_n does not appear in the positive-sequence circuit because the positive-sequence current represents balanced three-phase currents. Since the sum of such currents is equal to zero, they do not flow through the neutral connection. For the same reason, Z_n does not appear in the negative-sequence circuit.
- On the other hand, the zero-sequence impedance is $Z_0 = Z_Y + 3Z_n$ because the zero-sequence current represents three currents that have the same magnitude and phase. These currents add together at node n and must exit the load through the neutral connection. The factor 3 stems from the fact that $\overline{I}_n = 3\overline{I}_0$.
- If the neutral point n is not connected to ground, $Z_n = \infty$, and hence $Z_0 = \infty$. The zero-sequence network is then an open circuit, which means that a zero-sequence current can

flow only if there exists a path to ground. In particular, the zero-sequence current in a delta-connected load is always zero.

- If a balanced set of voltages in the positive phase sequence is applied to this load, $\overline{V_1} \neq 0$, while $\overline{V_0} = \overline{V_2} = 0$. Hence $\overline{I_1} \neq 0$, while $\overline{I_0} = \overline{I_2} = 0$. The currents flowing in this load are thus, as expected, balanced in the positive phase sequence. The positive-sequence network is thus equivalent to the single-phase representation of balanced three-phase systems that we introduced in Chapter 4.

It is important to note that the decoupling between the sequence networks illustrated in Figure 10.4 occurs because the impedances in the three phases are equal. If these impedances are not equal, the sequence networks are coupled. Applying a positive-sequence voltage to unbalanced impedances would result in negative-sequence currents and possibly zero-sequence currents.

Example 10.6 *Applying Unbalanced Voltages to a Y-Connected Load* Each branch of a Y-connected load has an impedance of $9 + j3\,\Omega$. The neutral point of this load is connected to ground through an impedance of $j2\,\Omega$. The following set of unbalanced line-to-neutral voltages is applied to this load:

$$\overline{V_a} = 100\angle 0° \text{ V}; \quad \overline{V_b} = 90\angle 240° \text{ V}; \quad \overline{V_c} = 110\angle 120° \text{ V}$$

Calculate the current in each phase and in the neutral connection.

Since the impedances in each phase are equal, (10.33)–(10.35) are valid and we have:

$$Z_0 = Z_Y + 3Z_n = 9 + j3 + 3 \times j2 = 9 + j9\,\Omega$$

$$Z_1 = Z_Y = 9 + j3 = 9 + j3\,\Omega$$

$$Z_2 = Z_Y = 9 + j3 = 9 + j3\,\Omega$$

We transformed these phase voltages into their zero-, positive-, and negative-sequence voltages in Example 10.3 and found that:

$$\overline{V_0} = 5.77\angle 90° \text{ V}$$

$$\overline{V_1} = 100\angle 0° \text{ V}$$

$$\overline{V_2} = 5.77\angle -90° \text{ V}$$

Hence:

$$\overline{I_0} = \frac{\overline{V_0}}{Z_0} = \frac{5.77\angle 90°}{9 + j9} = 0.453\angle 45° \text{ A}$$

$$\overline{I_1} = \frac{\overline{V_1}}{Z_1} = \frac{100\angle 0°}{9 + j3} = 10.54\angle -18.43° \text{ A}$$

$$\overline{I_2} = \frac{\overline{V_2}}{Z_3} = \frac{5.77\angle -90°}{9 + j3} = 0.608\angle -108.43° \text{ A}$$

Converting these sequence currents back into phase currents using (10.15), we get:

$$\overline{I_a} = 10.75 \angle -19.51° \text{ A}$$

$$\overline{I_b} = 9.57 \angle -136.77° \text{ A}$$

$$\overline{I_c} = 11.32 \angle +101.20° \text{ A}$$

$$\overline{I_n} = 3\overline{I_0} = 1.359 \angle 45° \text{ A}$$

Example 10.7 *Applying Unbalanced Voltages to a Delta-connected Load* Repeat the calculations of Example 10.5 assuming that the branch impedances are connected in delta.

We first need to replace the delta-connected load by its Y-connected equivalent.

$$Z_Y = \frac{Z_\Delta}{3} = \frac{9 + j3}{3} = 3 + j1\Omega$$

$Z_n = \infty$ because there is no neutral connection in a delta-connected load. Hence:

$$Z_0 = \infty$$

$$Z_1 = 3 + j1\Omega$$

$$Z_2 = 3 + j1\Omega$$

The currents in the sequence networks are:

$$\overline{I_0} = \frac{\overline{V_0}}{Z_0} = 0$$

$$\overline{I_1} = \frac{\overline{V_1}}{Z_1} = \frac{100 \angle 0°}{3 + j1} = 31.62 \angle -18.43° \text{ A}$$

$$\overline{I_2} = \frac{\overline{V_2}}{Z_3} = \frac{5.77 \angle -90°}{3 + j1} = 1.824 \angle -108.43° \text{ A}$$

Converting these values using (10.15), we get the line currents flowing into the load:

$$\overline{I_a} = 31.67 \angle -21.73° \text{ A}$$

$$\overline{I_b} = 30.05 \angle -136.69° \text{ A}$$

$$\overline{I_c} = 33.21 \angle +103.14° \text{ A}$$

10.3.2 Sequence Networks Representation of Generators

Figure 10.5 shows the circuit diagram of a Y-connected synchronous generator, whose neutral point is connected to ground through an impedance. To study the behavior of this generator under asymmetrical conditions, we need to convert this representation in terms of phase variables into positive-, negative-, and zero-sequence networks. Because the three phases of this generator are balanced, this conversion produces the three decoupled network shown in Figure 10.6.

Since the generator is designed to produce a balanced set of voltages in the positive phase sequence, the internal e.m.f. appears only in the positive-sequence network. Depending on the timeframe under consideration, the positive-sequence impedance is equal to the synchronous, transient, or subtransient reactance of this generator. Since the positive-sequence currents are balanced, they don't flow through the neutral connection and the neutral-to-ground impedance Z_n does not affect the positive-sequence impedance Z_1. The positive-sequence network is thus identical to the per-phase equivalent circuit that we introduced in Chapter 4.

The negative-sequence network consists only of the impedance to the negative-sequence currents Z_2 because the generator does not produce negative-sequence voltages. Since these currents create a flux wave that rotates in the opposite direction as the rotor, the

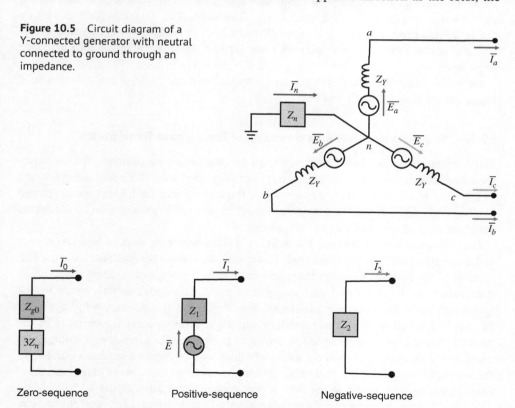

Figure 10.5 Circuit diagram of a Y-connected generator with neutral connected to ground through an impedance.

Zero-sequence

Positive-sequence

Negative-sequence

Figure 10.6 Sequence network representation of a synchronous generator.

damper windings prevent this flux wave from penetrating the rotor. The negative-sequence impedance Z_2 is thus smaller than the steady-state synchronous reactance of the generator. For fault studies, the positive-sequence reactance is taken as the subsynchronous reactance of the generator. Due to the effect of the damper windings, the positive-sequence impedance Z_1 is roughly equal to the negative-sequence impedance Z_1.

The zero-sequence network also does not include a voltage source. Since $\overline{I}_n = 3\overline{I}_0$, the impedance to the zero-sequence currents Z_0 consists of $3Z_n$ plus the impedance to these currents internal to the generator Z_{g0}. Zero-sequence currents can thus flow through a generator only if it is Y-connected and the neutral is connected to ground through a finite impedance.

For fault studies, the resistance of the windings is typically neglected.

10.3.3 Sequence Networks Representation of Three-phase Lines and Cables

If the conductors carrying the three phases of a transmission line are equidistant or if they are regularly transposed to approximate this equidistance, their impedances are equal, and the sequence networks are decoupled. Furthermore, the positive- and negative-sequence impedances are then equal because the phase sequence does not matter. Because the zero-sequence currents are identical in magnitude and phase in all conductors, their return path involves the ground or the overhead ground wires. The magnetic field created by the zero-sequence currents is thus quite different from the magnetic field created by the positive- and negative-sequence current, resulting in a significantly larger zero-sequence impedance.

We will assume that the resistance and shunt admittance of lines and cables have a negligible effect on fault calculations.

10.3.4 Sequence Networks Representation of Three-phase Transformers

Since transformers are indifferent to the phase sequence, the positive- and negative-sequence impedances have the same value. For simplicity, we will neglect the resistance of the windings and the shunt admittance in the transformer model that we developed in Chapter 7. The positive- and negative-sequence network representation thus consists simply of the leakage reactance of the transformer.

The zero-sequence network representation of a transformer depends on how its primary and secondary windings are connected. To explain the various possibilities, we must first recall that the currents in the primary and secondary windings of an ideal single-phase transformer are in phase and that the ratio of their magnitudes is determined by the turns-ratio. In a three-phase transformer, the currents in the primary windings must be matched by currents in the secondary windings. Positive-sequence currents in the primary windings thus require positive-sequence currents in the secondary windings, and negative-sequence currents in the primary windings require negative-sequence currents in the secondary windings. This is always possible. On the other hand, some combinations of winding connections preclude the flow of zero-sequence currents. Figure 10.7 illustrates five combinations of primary and secondary windings connections. For each connection, it shows from left to right how the connections would be displayed on a one-line diagram,

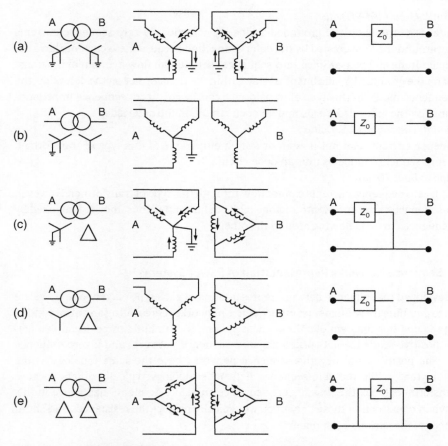

Figure 10.7 Zero-sequence network representation for different transformer connections: (a) grounded Y–grounded Y; (b) grounded Y–ungrounded Y; (c) grounded Y–delta; (d) ungrounded Y–delta; (e) delta–delta.

a schematic representation of how the windings are connected and how zero-sequence currents would flow, and the corresponding zero-sequence network.

Connection a: Grounded-Y/Grounded-Y

Because the neutral is connected to ground on both sides, zero-sequence currents can flow in and out of this transformer. Moreover, as the arrows in the middle diagram illustrate, the primary and secondary currents in each phase or the transformer are matched. The zero-sequence network representation of this type of transformer thus consists of a series impedance between the primary and secondary sides.

Connection b: Grounded-Y/Ungrounded-Y

While it is tempting to think that zero-sequence currents could flow on the side where the neutral is grounded, such currents would not be matched by zero-sequence currents on the ungrounded side. This type of transformer thus blocks the flow of zero-sequence currents.

Connection c: Grounded-Y/Delta

Zero-sequence currents on the grounded-Y side are matched by zero-sequence currents flowing around the loop created by the delta connection because these currents have the same magnitude and phase. Since zero-sequence currents can flow in or out of the transformer on the grounded-Y side but must stay within the transformer on the delta side, the zero-sequence model of this type of transformer consists of a zero-sequence impedance to ground on the grounded-Y side and an open circuit on the delta side.

Connection d: Ungrounded-Y/Delta

Zero-sequence currents can not flow in or out on either side of this type of transformer. Their zero-sequence model is thus an open circuit.

Connection e: Delta/Delta

Here again, zero-sequence currents cannot flow through this type of transformer. However, if zero-sequence currents flow around one of the delta connections, they are matched by zero-sequence currents flowing around the other.

10.3.5 Sequence Networks Representation of Power Systems

Having developed the sequence network representation of each type of component, we are now able to combine these elements into sequence networks representing an entire system. Based on the one-line diagram of a three-phase system, we can build three decoupled networks: A positive-sequence network, a negative-sequence network, and a zero-sequence network. The positive- and negative-sequence networks have the same topology as the one-line diagram, while the zero-sequence network's topology typically differs because some transformers block the flow of zero-sequence currents. As mentioned above in this section, when constructing these networks, we typically neglect the resistances and shunt admittances of the various components.

Example 10.8 *Sequence Networks* Figure 10.8 shows the one-line diagram of the small power system that we considered in Example 9.2. Symbols indicate the connections of the windings of the generators and transformers. Table 10.1 gives the values of the positive-, negative-, and zero-sequence reactances of each component in per unit on a consistent basis.

Figure 10.9 displays the sequence network representation of this system. The positive-sequence network is identical to the single-phase network representation of Figure 9.3. If the system is unloaded and operating at nominal voltage, all the positive-sequence

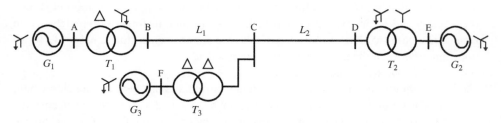

Figure 10.8 One-line diagram of the system for the examples on unbalanced faults.

Table 10.1 Positive-, negative-, and zero-sequence reactances of the components of the system of Example 10.8.

	X_1 (pu)	X_2 (pu)	X_0 (pu)
G_1	0.5	0.2	0.18
G_2	0.33	0.15	0.15
G_3	0.275	0.1	0.12
T_1	0.2	0.2	0.2
T_2	0.167	0.167	0.167
T_3	0.143	0.143	0.143
L_1	0.083	0.083	0.4
L_2	0.103	0.103	0.6

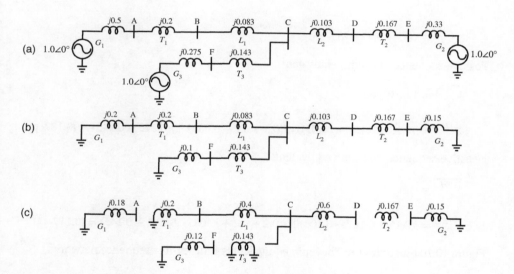

Figure 10.9 Sequence networks representation of the system for the examples on unbalanced faults (a) positive-sequence; (b) negative-sequence; (c) zero-sequence. The values indicated are impedances in per unit.

voltage sources have a value of $1.0 \angle 0°$ pu. The negative-sequence network has the same topology as the positive-sequence network but without voltage sources because we assume that the voltages are balanced. To determine the topology of the zero-sequence network, we must consider the connections of the transformers and generators. Since the neutrals of G_1 and G_3 are connected to ground, so is their zero-sequence impedance. On the other hand, the zero-sequence impedance of G_2 is open-ended because the neutral of that generator is ungrounded. Because Transformer T_1 is connected in the delta/grounded-Y configuration, its zero-sequence network is open-ended on the side of node A. Since transformer T_2 has a grounded-Y connection on both sides, its zero-sequence impedance connects nodes D and E. Finally, since both the primary and secondary of transformer T_3 are connected in delta, its zero-sequence is connected to ground on both sides. There are no voltage sources in the zero-sequence network either. Because the system is assumed to be balanced before a fault occurs, only the positive-sequence network is "live," and these three networks are decoupled. In the Section 10.4, we will see how unbalanced faults connect these networks.

Example 10.9 *Thevenin Equivalents of Sequence Networks* All the standard network analysis techniques are applicable to these sequence networks. For example, let us determine the Thevenin equivalents as seen from node C of the sequence networks of Figure 10.9. Assuming the system is balanced, unloaded, and operating at nominal voltage, we have:

Zero-sequence Thevenin equivalent:

$$\overline{V_C^{TH,0}} = 0$$

$$Z_C^{TH,0} = j(0.2 + 0.4) = j0.6 \text{ pu}$$

Positive-sequence Thevenin equivalent:

$$\overline{V_C^{TH,1}} = 1.0\angle0° \text{ pu}$$

$$Z_C^{TH,1} = j(0.33 + 0.167 + 0.103) \parallel j(0.5 + 0.2 + 0.083) \parallel j(0.275 + 0.143) = j0.187 \text{ pu}$$

Negative-sequence Thevenin equivalent:

$$\overline{V_C^{TH,2}} = 0$$

$$Z_C^{TH,2} = j(0.15 + 0.167 + 0.103) \parallel j(0.2 + 0.2 + j0.083) \parallel j(0.1 + 0.143) = j0.117 \text{ pu}$$

Figure 10.10 illustrates the Thevenin equivalents of these three sequence networks.

10.4 Unbalanced Faults

Until the occurrence of a fault, the power network is a closed system where the currents circulate on established paths. To underline this point in Figure 10.11a, the whole network is shown as circumscribed by a closed contour. Suppose that we are interested in figuring out what would happen if an unbalanced fault were to occur at a particular node k. By definition, a fault creates a path for currents to flow out the network at node k and then back in. To calculate these currents, we must model the network as seen for this node. Instead of using the three-phase network representation shown in Figure 10.11a, we can use the Thevenin network equivalents from node k of its sequence networks, as shown in Figure 10.11b.

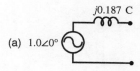

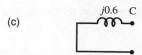

Figure 10.10 Thevenin equivalent of the sequence networks representation of the system for the examples on unbalanced faults as seen from node C, (a) positive-sequence; (b) negative-sequence; (c) zero-sequence.

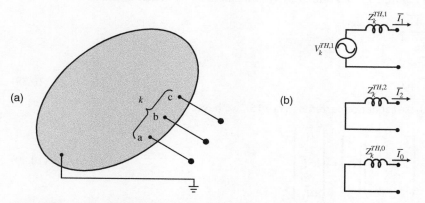

Figure 10.11 (a) Conceptual three-phase representation of a network for unbalanced fault calculations at a particular node k. (b) Representation of this network by the Thevenin equivalents of its sequence networks.

10.4.1 Balanced Three-phase Fault

Let us begin by showing that using sequence networks to analyze balanced faults gives the same results as those obtained in Chapter 9. Suppose as illustrated in Figure 10.12a that a solid, balanced, three-phase fault occurs at node k of the network. Since such a fault can be viewed as connecting all three phases to the neutral point through a zero impedance, we have:

$$\overline{V}_a = \overline{V}_b = \overline{V}_c = 0 \tag{10.36}$$

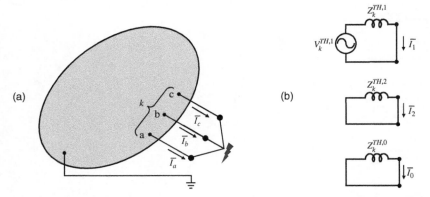

Figure 10.12 Balanced three-phase fault at node k. (a) Three-phase representation; (b) Sequence network representation.

Inserting these voltage values in (10.13), we get:

$$\overline{V_0} = \overline{V_1} = \overline{V_2} = 0 \tag{10.37}$$

The Thevenin equivalents of the three sequence networks are thus short-circuited, as illustrated in Figure 10.12b. From these circuit diagrams we conclude that:

$$\overline{I_1} = \frac{V_1}{Z_k^{TH,1}}$$

$$\overline{I_2} = 0$$

$$\overline{I_0} = 0 \tag{10.38}$$

Inserting these sequence currents in (10.15), we have:

$$\begin{pmatrix} \overline{I_a} \\ \overline{I_b} \\ \overline{I_c} \end{pmatrix} = \begin{pmatrix} 1 & 1 & 1 \\ 1 & a^2 & a \\ 1 & a & a^2 \end{pmatrix} \begin{pmatrix} 0 \\ \overline{I_1} \\ 0 \end{pmatrix} = \begin{pmatrix} \overline{I_1} \\ a^2\overline{I_1} \\ a\overline{I_1} \end{pmatrix} \tag{10.39}$$

The fault currents are thus balanced, and their magnitude is limited by the Thevenin equivalent of the positive-sequence network. The positive-sequence network is thus equivalent to the per-phase network used to study balanced faults in Chapter 9.

10.4.2 Single Line-to-ground Fault

Let us now consider the case of a bolted or solid fault between a single phase and ground at node k of the network. Because the labeling of the phases is arbitrary, we can assume without loss of generality and as illustrated in Figure 10.13a, that this fault occurs between phase a and ground. The fault current therefore flows only in phase a:

$$\overline{I_a} = \overline{I_F} \neq 0$$

$$\overline{I_b} = \overline{I_c} = 0 \tag{10.40}$$

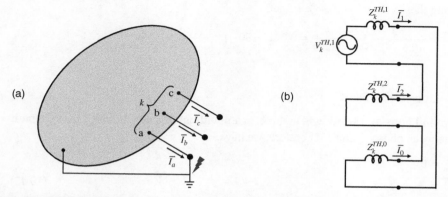

Figure 10.13 Solid line-to-ground fault at node k. (a) Three-phase representation; (b) Sequence network representation.

To convert these phase currents into their symmetrical components, we insert these values in (10.16):

$$\begin{pmatrix} \overline{I}_0 \\ \overline{I}_1 \\ \overline{I}_2 \end{pmatrix} = \frac{1}{3} \begin{pmatrix} 1 & 1 & 1 \\ 1 & a & a^2 \\ 1 & a^2 & a \end{pmatrix} \begin{pmatrix} \overline{I}_F \\ 0 \\ 0 \end{pmatrix}$$

(10.41)

Hence:

$$\overline{I}_0 = \overline{I}_1 = \overline{I}_2 = \frac{1}{3}\overline{I}_F$$

(10.42)

A single-phase to ground fault is thus equivalent to forcing the positive-, negative-, and zero-sequence currents to be equal. As Figure 10.13b illustrates, we achieve this by connecting in series the positive-, negative-, and zero-sequence circuits at the node where the fault occurs. The current in this series combination is:

$$\overline{I}_0 = \overline{I}_1 = \overline{I}_2 = \frac{V_k^{TH,1}}{Z_k^{TH,1} + Z_k^{TH,2} + Z_k^{TH,0}}$$

(10.43)

Combining (10.42) and (10.43), we get the single-phase fault current:

$$\overline{I}_F = \frac{3V_k^{TH,1}}{Z_k^{TH,1} + Z_k^{TH,2} + Z_k^{TH,0}}$$

(10.44)

Example 10.10   *Single Line-to-ground Fault*   Let us calculate the current that would result from a single-phase to ground fault at node C of Figure 10.8, assuming that the system is unloaded and operating at nominal voltage. In Example 10.8, we calculated the Thevenin equivalents of the positive-, negative-, and zero-sequence networks as seen from this node:

$$\overline{V}_C^{TH,1} = 1.0\angle 0° \text{ pu}$$

$$Z_C^{TH,0} = j0.6 \text{ pu}$$

$$Z_C^{TH,1} = j0.187 \text{ pu}$$

$$Z_C^{TH,2} = j0.117 \text{ pu}$$

Inserting these values in (10.44), we get:

$$\overline{I_F} = \frac{3 \times 1.0 \angle 0°}{j(0.6 + 0.187 + 0.117)} = 3.32 \angle -90° \text{ pu}$$

10.4.3 Line-to-line Fault

As shown in Figure 10.14a, we will assume, again without loss of generality, that the fault occurs between phases b and c. Therefore, we have:

$$\overline{I_a} = 0$$
$$\overline{I_b} = -\overline{I_c} = \overline{I_F} \neq 0 \tag{10.45}$$

We again insert these values in (10.16) to model this type of fault in terms of symmetrical components:

$$\begin{pmatrix} \overline{I_0} \\ \overline{I_1} \\ \overline{I_2} \end{pmatrix} = \frac{1}{3} \begin{pmatrix} 1 & 1 & 1 \\ 1 & a & a^2 \\ 1 & a^2 & a \end{pmatrix} \begin{pmatrix} 0 \\ \overline{I_F} \\ -\overline{I_F} \end{pmatrix} \tag{10.46}$$

Hence:

$$\overline{I_0} = \frac{1}{3}(\overline{I_F} - \overline{I_F}) = 0$$
$$\overline{I_1} = \frac{1}{3}(a - a^2)\overline{I_F}$$
$$\overline{I_2} = \frac{1}{3}(a^2 - a)\overline{I_F} = -\overline{I_1} \tag{10.47}$$

A line-to-line fault therefore does not create a zero-sequence current. To reflect the fact that the positive- and negative-sequence currents flow in opposite direction, we connect their sequence networks in anti-parallel, as shown in Figure 10.14b. The values of these currents are thus:

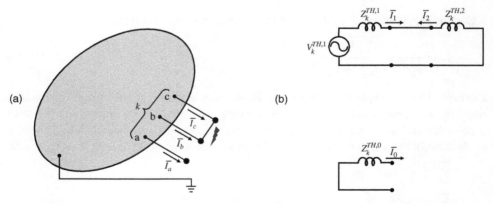

(a)

(b)

Figure 10.14 Solid line-to-line fault. (a) Three-phase representation; (b) Sequence network representation.

$$\overline{I}_1 = -\overline{I}_2 = \frac{\overline{V_k^{TH,1}}}{Z_k^{TH,1} + Z_k^{TH,2}} \tag{10.48}$$

Since $a - a^2 = 1\angle120° - 1\angle240° = j\sqrt{3}$, the line-to-line fault current is:

$$\overline{I}_F = \frac{3}{j\sqrt{3}}\overline{I}_1 = -j\sqrt{3} \times \frac{\overline{V_k^{TH,1}}}{Z_k^{TH,1} + Z_k^{TH,2}} \tag{10.49}$$

Example 10.11 *Line-to-line Fault* Considering again the system of Figure 10.8, let us calculate the line-to-line current that would result from a fault at node C. Assuming that the system is unloaded and operating at nominal voltage and using the Thevenin impedances calculated in Example 10.8, we have:

$$\overline{I}_F = \frac{3}{j\sqrt{3}}\overline{I}_1 = -j\sqrt{3} \times \frac{1.0\angle0°}{j(0.187 + 0.117)} = -5.70\angle0° \text{ pu}$$

10.4.4 Double Line-to-ground Fault

This type of fault creates a zero-impedance path between two phases and ground. If we assume as shown in Figure 10.15a that it involves phases b and c, we have:

$$\overline{V}_b = \overline{V}_c = 0 \tag{10.50}$$

Inserting the voltages in (10.13), we get:

$$\begin{pmatrix} \overline{V}_0 \\ \overline{V}_1 \\ \overline{V}_2 \end{pmatrix} = \frac{1}{3}\begin{pmatrix} 1 & 1 & 1 \\ 1 & a & a^2 \\ 1 & a^2 & a \end{pmatrix}\begin{pmatrix} \overline{V}_a \\ 0 \\ 0 \end{pmatrix} = \frac{1}{3}\begin{pmatrix} \overline{V}_a \\ \overline{V}_a \\ \overline{V}_a \end{pmatrix} \tag{10.51}$$

Or:

$$\overline{V}_0 = \overline{V}_1 = \overline{V}_2 = \frac{1}{3}\overline{V}_a \tag{10.52}$$

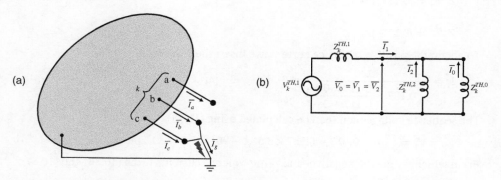

Figure 10.15 Double line-to-ground fault. (a) Three-phase representation; (b) Sequence network representation.

Since phase a is not faulted, $\overline{I}_a = 0$ and the first row of (10.15) gives:

$$\overline{I}_0 + \overline{I}_1 + \overline{I}_2 = 0 \tag{10.53}$$

Equations (10.52) and (10.53) suggest that we can study double line-to-ground faults by connecting the Thevenin equivalent of the three sequence networks in parallel at node k where the fault occurs, as illustrated in Figure 10.15.b. From this circuit diagram we can derive the following expressions:

$$\overline{I}_1 = \frac{\overline{V_k^{TH,1}}}{Z_k^{TH,1} + \dfrac{Z_k^{TH,2} Z_k^{TH,0}}{Z_k^{TH,2} Z_k^{TH,0}}} \tag{10.54}$$

$$\overline{V}_0 = \overline{V}_1 = \overline{V}_2 = \overline{V_k^{TH,1}} - Z_k^{TH,1}\overline{I}_1 \tag{10.55}$$

$$\overline{I}_2 = -\frac{\overline{V}_2}{Z_k^{TH,2}} \tag{10.56}$$

$$\overline{I}_0 = -\frac{\overline{V}_0}{Z_k^{TH,0}} \tag{10.57}$$

Once the values of $\overline{I}_0$, $\overline{I}_1$, and $\overline{I}_2$ have been calculated, they can be inserted in (10.15) to determine $\overline{I}_b$ and $\overline{I}_c$. The current in the ground connection is then:

$$\overline{I}_g = \overline{I}_b + \overline{I}_c \tag{10.58}$$

Example 10.12 *Double Line-to-ground Fault*   Let us calculate the currents flowing as a result of a double line-to-ground fault at node C in the system of Figure 10.8. We will again assume that the system is unloaded and that the pre-fault voltages are at nominal value. Using the Thevenin equivalents calculated in Example 10.8, we have:

$$\overline{V_C^{TH,1}} = 1.0\angle 0° \text{ pu}$$

$$Z_C^{TH,0} = j0.6 \text{ pu}$$

$$Z_C^{TH,1} = j0.187 \text{ pu}$$

$$Z_C^{TH,2} = j0.117 \text{ pu}$$

To obtain the positive-sequence current, we insert these values in (10.54):

$$\overline{I}_1 = \frac{1.0\angle 0°}{j\left(0.187 + \dfrac{0.117 \times 0.6}{0.117 + 0.6}\right)} = 3.51\angle -90° \text{ pu}$$

The sequence voltages can then be calculated using (10.55):

$$\overline{V}_0 = \overline{V}_1 = \overline{V}_2 = 1.0\angle 0° - j0.187 \times 3.51\angle -90° = 0.343\angle 0° \text{ pu}$$

From which we get the negative- and zero-sequence currents using (10.56) and (10.57):

$$\overline{I}_2 = -\frac{0.343\angle 0°}{j0.117} = 2.93\angle 90° \text{ pu}$$

$$\overline{I}_0 = -\frac{0.343\angle 0°}{j0.6} = 0.572\angle 90° \ \text{pu}$$

To obtain the phase currents, we insert the sequence currents in (10.15):

$$\begin{pmatrix} \overline{I}_a \\ \overline{I}_b \\ \overline{I}_c \end{pmatrix} = \begin{pmatrix} 1 & 1 & 1 \\ 1 & a^2 & a \\ 1 & a & a^2 \end{pmatrix} \begin{pmatrix} 0.572\angle 90° \\ 3.51\angle -90° \\ 2.93\angle 90° \end{pmatrix} = \begin{pmatrix} 0 \\ 5.64\angle 171.21° \\ 5.64\angle 8.79° \end{pmatrix} \ \text{pu}$$

The current in the ground connection is then:

$$\overline{I}_g = 5.64\angle 171.21° + 5.64\angle 8.79° = 1.72\angle 90° \ \text{pu}$$

10.5 Unbalanced Fault Calculations in Large Systems

Because the networks of the previous examples are small, we were able to calculate the Thevenin equivalents by hand. Since this is clearly not feasible for any practical power system, we need to generalize the method described in Section 9.3 to handle unbalanced faults. In that section, we showed that the Thevenin impedance needed to calculate the current resulting from a balanced fault at node k was the (k,k) term of the impedance matrix of the network. Because we were dealing with balanced faults, this impedance matrix represented the positive-sequence network. To handle unbalanced faults, we need to introduce the zero-sequence and negative-sequence impedance matrices. To distinguish these matrices, we assign them the superscript of their sequence network. The Thevenin equivalent impedances of the sequence networks as seen from node k are thus:

$$Z_k^{TH,0} = Z_0(k,k)$$
$$Z_k^{TH,1} = Z_1(k,k)$$
$$Z_k^{TH,2} = Z_2(k,k) \tag{10.59}$$

These values can then be used in the formulas derived in the Sections 10.4.1–10.4.4 to calculate the currents resulting from unbalanced faults. For example, based on (10.44) the fault current resulting from a single line-to-ground fault at node k is given by:

$$\overline{I}_F = \frac{3\overline{V}_k^{TH,1}}{Z_1(k,k) + Z_2(k,k) + Z_0(k,k)} \tag{10.60}$$

As illustrated in the following example, to obtain these impedance matrices, we build and invert the admittance matrices of their respective sequence network.

Example 10.13 *Impedance Matrices of Sequence Networks* Consider the small system of Figure 10.8. To help us build its admittance matrices, Figure 10.16 shows its positive-, negative-, and zero-sequence networks representations in terms of admittances. Numbering the nodes in alphabetical order, we build the three admittance matrices by inspecting these circuit diagrams.

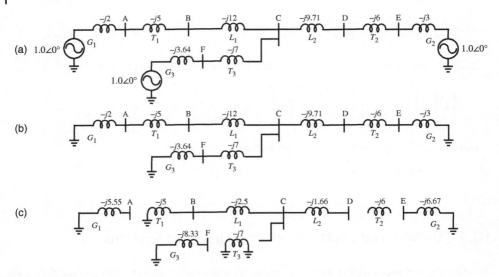

Figure 10.16 Sequence networks representation of the system of Example 10.13. (a) Positive-sequence; (b) negative-sequence; (c) zero-sequence. The values shown are admittances in per unit.

Positive-sequence admittance matrix:

$$Y_1 = j \begin{pmatrix} -7 & 5 & 0 & 0 & 0 & 0 \\ 5 & -17 & 12 & 0 & 0 & 0 \\ 0 & 12 & -28.71 & 9.71 & 0 & 7 \\ 0 & 0 & 9.71 & -15.71 & 6 & 0 \\ 0 & 0 & 0 & 6 & -9 & 0 \\ 0 & 0 & 7 & 0 & 0 & -10.64 \end{pmatrix} \text{pu}$$

Negative-sequence admittance matrix:

$$Y_2 = j \begin{pmatrix} -7 & 5 & 0 & 0 & 0 & 0 \\ 5 & -17 & 12 & 0 & 0 & 0 \\ 0 & 12 & -28.75 & 9.71 & 0 & 7 \\ 0 & 0 & 9.71 & -15.71 & 6 & 0 \\ 0 & 0 & 0 & 6 & -12.66 & 0 \\ 0 & 0 & 7 & 0 & 0 & -17 \end{pmatrix} \text{pu}$$

Zero-sequence admittance matrix:

$$Y_0 = j \begin{pmatrix} -5.55 & 0 & 0 & 0 & 0 & 0 \\ 0 & -7.5 & 2.5 & 0 & 0 & 0 \\ 0 & 2.5 & -4.16 & 1.66 & 0 & 0 \\ 0 & 0 & 1.66 & -1.66 & 0 & 0 \\ 0 & 0 & 0 & 0 & -6.67 & 0 \\ 0 & 0 & 0 & 0 & 0 & -8.33 \end{pmatrix} \text{pu}$$

Note that the difference between the structure of the zero-sequence admittance matrix and the structure of the other two. This difference stems from the difference in the topologies of these networks.

Inverting these matrices, we get:
Positive-sequence impedance matrix:

$$
Z_1 = j \begin{pmatrix}
0.257 & 0.160 & 0.120 & 0.099 & 0.066 & 0.079 \\
0.160 & 0.224 & 0.168 & 0.139 & 0.093 & 0.110 \\
0.120 & 0.168 & 0.187 & 0.156 & 0.104 & 0.123 \\
0.099 & 0.139 & 0.156 & 0.214 & 0.143 & 0.102 \\
0.066 & 0.093 & 0.104 & 0.143 & 0.206 & 0.068 \\
0.079 & 0.110 & 0.123 & 0.102 & 0.068 & 0.175
\end{pmatrix} \text{pu}
$$

Negative-sequence impedance matrix:

$$
Z_2 = j \begin{pmatrix}
0.137 & 0.074 & 0.048 & 0.036 & 0.017 & 0.020 \\
0.074 & 0.149 & 0.096 & 0.073 & 0.034 & 0.040 \\
0.048 & 0.096 & 0.116 & 0.088 & 0.042 & 0.048 \\
0.036 & 0.073 & 0.088 & 0.144 & 0.068 & 0.036 \\
0.017 & 0.034 & 0.042 & 0.068 & 0.111 & 0.017 \\
0.020 & 0.040 & 0.048 & 0.036 & 0.017 & 0.079
\end{pmatrix} \text{pu}
$$

Zero-sequence impedance matrix:

$$
Z_0 = j \begin{pmatrix}
0.180 & 0.000 & 0.000 & 0.000 & 0.000 & 0.000 \\
0.000 & 0.200 & 0.200 & 0.200 & 0.000 & 0.000 \\
0.000 & 0.200 & 0.600 & 0.600 & 0.000 & 0.000 \\
0.000 & 0.200 & 0.600 & 1.202 & 0.000 & 0.000 \\
0.000 & 0.000 & 0.000 & 0.000 & 1.500 & 0.000 \\
0.000 & 0.000 & 0.000 & 0.000 & 0.000 & 0.120
\end{pmatrix} \text{pu}
$$

Example 10.14 *Unbalanced Faults in Large Networks* Using the results from the previous example, we can easily calculate the currents resulting from balanced or unbalanced faults at all the bus of the system of Figure 10.8. Table 10.2 shows the magnitude of the single-phase to ground and three-phase fault currents. Line-to-line and double line-to-ground faults have not been included because they cause smaller fault currents

Table 10.2 Magnitude of the fault currents resulting from single-phase and three-phase faults at every node of the system of Figure 10.8.

Node	$Z_0(k, k)$ (pu)	$Z_1(k, k)$ (pu)	$Z_2(k, k)$ (pu)	Single-phase fault (pu)	Three-phase fault (pu)
A	0.18	0.257	0.137	5.226	3.891
B	0.2	0.224	0.149	5.236	4.464
C	0.6	0.187	0.116	3.322	5.348
D	1.202	0.214	0.144	1.923	4.673
E	1.500	0.206	0.111	1.651	4.854
F	0.120	0.175	0.079	8.021	5.714

than the other types. While three-phase faults are usually the most severe, in locations where the zero-sequence Thevenin impedance is small, such as nodes A, B, and F in this example, a single-phase fault produces a larger fault current than a three-phase fault.

References

Fortescue, C.L. (1918). Method of symmetrical co-ordinates applied to the solution of polyphase networks. *Transactions of the American Institute of Electrical Engineers* XXXVII (2): 1027–1140. https://doi.org/10.1109/T-AIEE.1918.4765570.

Kreysig, E. (2015). *Advanced Engineering Mathematics*, 10e. Wiley.

Further Reading

Anderson, P.M. (1995). *Analysis of Faulted Power Systems*. Wiley-IEEE Press. This classic book provides a detailed treatment of the various types of faults.

Power Systems Modelling and Fault Analysis: Theory and Practice, 2, by Nasser Tleis, Academic Press, 2019.

Power System Analysis by John Grainger, William Stevenson, McGraw Hill, 1994.

Power System Analysis and Design, 6, by J. Duncan Glover, Thomas Overbye, Mulukutla S. Sarma, Cengage Learning, 2016.

Problems

P10.1 Given $a = 1\angle 120° = -0.5 + j0.866$, calculate the following: a^4; a^5; a^6; $a - a^2$; $1 - a$.

P10.2 Given $a = 1\angle 120°$, show that $1 + a + a^2 = 0$; $\frac{1}{a} = a^2$; $\frac{1-a}{a-a^2} = a^2$; $\frac{a^2-1}{a-a^2} = a$

Hint: Except for the first two, you do not need to replace a by its value in rectangular or polar coordinates.

P10.3 Given the matrix $A = \begin{pmatrix} 1 & 1 & 1 \\ 1 & a^2 & a \\ 1 & a & a^2 \end{pmatrix}$ with $a = 1\angle 120°$, calculate its inverse A^{-1}.

Hints:

1. Use the following formula to carry out the matrix inversion:

$$A^{-1} = \frac{1}{\det A} \begin{pmatrix} A_{11} & A_{21} & A_{31} \\ A_{12} & A_{22} & A_{32} \\ A_{13} & A_{23} & A_{33} \end{pmatrix}$$ where A_{ij} is the cofactor of element a_{ij} of matrix A.

(For details, see for example Kreysig (2015).)

2. Do not replace a by its value in rectangular or polar coordinates. Use instead the formulas demonstrated in the previous problem.

P10.4 Write two computer programs: one to convert phase currents or voltages into their sequence components and the other to convert sequence components into phase quantities. Test these programs using the examples in the chapter. These programs will greatly facilitate the solution of the following problems.

P10.5 Calculate the sequence components of the following sets of phase values:

(a) $\overline{V_a} = 120\angle 90°$ V; $\overline{V_b} = 120\angle 330°$ V; $\overline{V_c} = 120\angle 210°$ V

(b) $\overline{I_a} = 240\angle 30°$ A; $\overline{I_b} = 240\angle 150°$ A; $\overline{I_c} = 240\angle 270°$ A

(c) $\overline{V_a} = 200\angle 0°$ V; $\overline{V_b} = 240\angle 240°$ V; $\overline{V_c} = 180\angle 120°$ V

(d) $\overline{I_a} = 120\angle 0°$ A; $\overline{I_b} = 120\angle 230°$ A; $\overline{I_c} = 120\angle 110°$ A

P10.6 Calculate the phase values corresponding to the following sequence components:

(a) $\overline{I_0} = 0\angle 0°$ A; $\overline{I_1} = 120\angle -30°$ A; $\overline{I_2} = 0\angle 0°$ A

(b) $\overline{V_0} = 0\angle 0°$ V; $\overline{V_1} = 0\angle 0°$ V; $\overline{V_2} = 100\angle 20°$ V

(c) $\overline{I_0} = 20\angle 15°$ A; $\overline{I_1} = 0\angle 0°$ A; $\overline{I_2} = 0\angle 0°$ A

(d) $\overline{V_0} = 0\angle 0°$ V; $\overline{V_1} = 240\angle 0°$ V; $\overline{V_2} = 20\angle 0°$ V

P10.7 A balanced Y-connected load consisting of an impedance of $(6+j3)\Omega$ in each phase is supplied from the following set of unbalanced line-to-ground voltages:

$$\overline{V_{ag}} = 300\angle 0°\ V$$

$$\overline{V_{bg}} = 280\angle -105°\ V$$

$$\overline{V_{cg}} = 320\angle 140°\ V$$

Calculate the line currents for the following three cases:
(a) The load neutral is solidly grounded.
(b) The load neutral is connected to ground through an inductive reactance of 2Ω.
(c) The connection between the load neutral and ground is open.

P10.8 A load consists of three $12+j6\Omega$ impedances connected in delta. The line currents flowing into this load are:

$$\overline{I_a} = 23.95\angle -1.66°\ A$$

$$\overline{I_b} = 17.70\angle -129.97°\ A$$

$$\overline{I_a} = 19.01\angle 131.41°\ A$$

Calculate the set of unbalanced line-to-neutral voltages applied to this load.

P10.9 The following set of unbalanced line-to-neutral voltages is applied to a balanced load:

$$\overline{V_{an}} = 278.19\angle 14.81°\ V$$

$$\overline{V}_{bn} = 125.29\angle - 121.44° \text{ V}$$

$$\overline{V}_{cn} = 212.28\angle 134.19° \text{ V}$$

The following line currents flow into this load:

$$\overline{I}_a = 54.38\angle - 2.69° \text{ A}$$

$$\overline{I}_b = 29.24\angle - 125.10° \text{ A}$$

$$\overline{I}_c = 36.91\angle - 128.02° \text{ A}$$

Determine the configuration of this load and the impedance of each branch.

P10.10 Show that:

$$\frac{1}{3}\begin{pmatrix} 1 & 1 & 1 \\ 1 & a & a^2 \\ 1 & a^2 & a \end{pmatrix}\begin{pmatrix} Z_Y + Z_n & Z_n & Z_n \\ Z_n & Z_Y + Z_n & Z_n \\ Z_n & Z_n & Z_Y + Z_n \end{pmatrix}\begin{pmatrix} 1 & 1 & 1 \\ 1 & a^2 & a \\ 1 & a & a^2 \end{pmatrix} = \begin{pmatrix} Z_Y + 3Z_n & 0 & 0 \\ 0 & Z_Y & 0 \\ 0 & 0 & Z_Y \end{pmatrix}$$

P10.11 Consider the small system shown in Figure P10.11. The table below gives the reactances of the components of this system in per unit on the basis of their rating for the generators and transformers and in Ohms for the lines.

	MVA rating	X_0	X_1	X_2
Generator G_1	40	0.05	0.15	0.20
Generator G_2	30	0.1	0.2	0.21
Transformer T_1	50	0.1	0.1	0.1
Transformer T_2	50	0.1	0.1	0.1
Line L_1	—	60 Ω	20 Ω	20 Ω
Line L_2	—	60 Ω	20 Ω	20 Ω

(a) Using a 100 MVA basis, convert these reactances to a consistent per unit basis.
(b) Draw the positive-, negative-, and zero-sequence network impedance diagrams of this system.

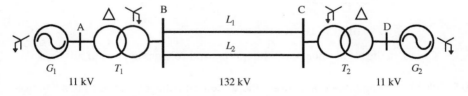

Figure P10.11 One-line diagram of the system of Problems P10.11–P10.17.

P10.12 Using the impedance diagrams that you developed in Problem P10.11, calculate the Thevenin equivalent of the positive-, negative-, and zero-sequence networks as seen from each node of the system. Assume that the system is operating at no-load and nominal voltage.

P10.13 Using the Thevenin equivalents that you calculated in Problem P10.12, calculate the current that would result from a balanced three-phase fault at each node of the system. Convert these values from per unit to Amperes.

P10.14 Using the Thevenin equivalents that you calculated in Problem P10.12, calculate the current that would result from a single-phase fault at each node of the system. Convert these values from per unit to Amperes.

P10.15 Using the Thevenin equivalents that you calculated in Problem P10.12, calculate the current that would result from a line-to-line fault at each node of the system. Convert these values from per unit to Amperes.

P10.16 Using the Thevenin equivalents that you calculated in Problem P10.12, calculate the current that would result from a double line-to-ground fault at node B of the system. Convert this value from per unit to Amperes.

P10.17 Using the impedance diagrams that you developed in Problem P10.11, build the admittance matrices of the positive-, negative-, and zero-sequence networks of this system. Invert these matrices to obtain the corresponding impedance matrices. Use these impedance matrices to check the results that you obtained in Problem P10.12.

P10.18 Consider the one-line diagram of a small power shown in Figure P10.18. The parameters of the components of this system are given on a consistent per unit basis in the table below. The circuit breaker CB connecting buses B and D is open. The neutral of G_2 is connected to ground through a reactance of 0.1 pu. The neutrals of the primary and secondary of transformer T_2 are solidly connected to ground. Assume that the system operates at nominal voltage and that the load currents are negligible compared to the fault currents.

	X_0 (pu)	X_1 (pu)	X_2 (pu)
G_1	0.05	0.2	0.2
G_2	0.03	0.2	0.1
T_1	0.1	0.1	0.1
T_2	0.15	0.15	0.15
L_1	1.8	0.6	0.6
L_2	1.8	0.6	0.6

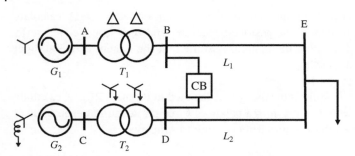

Figure P10.18 One-line diagram of the system of Problems P10.18 and P10.19.

(a) Draw the impedance and admittance diagrams of the positive-, negative-, and zero-sequence networks of this system.

(b) Calculate the positive-, negative-, and zero-sequence impedance matrices of this system.

(c) Calculate the per-unit fault current that would result from a single-line-to-ground fault at each of the buses in this system.

P10.19 Repeat the calculations of Problem P10.18 for the case where the circuit breaker CB connecting buses B and D is closed. Compare the results with those of Problem P10.18.

11

Introduction to Power System Stability

11.1 Overview

Over the course of a day the conditions under which a power system operates change. The loads fluctuate as a function of the aggregated consumers' demand and the generation pattern varies depending on the availability of renewable resources and the need to maintain the overall load/generation balance. Occasionally, the topology of the system changes as generating units and lines are switched on or off. If all goes well, the state of the system, characterized by the voltages and flows, varies smoothly as loads and generations evolve slowly within the expected range. Under these conditions, we say that the system is in a quasi-steady state because its equilibrium state shifts slowly over time. Given the load, generation, and topology of the system anticipated at a certain time, a power flow calculation reveals this equilibrium state. Therefore, as long as the power system remains in this quasi-steady state, we can track its evolution by performing power flow calculations based on a sequence of snapshots of loads, generation, and topology.

Occasionally, things go wrong. Loads can increase beyond the expected range. Generators occasionally fail and suddenly go off-line. Renewable energy sources can unexpectedly dwindle. Faults can trigger the disconnection of lines. Human errors happen. All such events can trigger rapid changes in the state of the system. A well-operated power system typically recovers from such mishaps and reaches a new equilibrium state. However, some events can trigger an instability or a cascading series of outages, leading to a partial or total collapse of the system.

In this chapter, we introduce some of the mechanisms that can lead to instabilities in power systems. However, a detailed treatment of the vast and complex topic of power system dynamics and stability is outside the scope of this book. The interested reader is therefore encouraged to consult the references listed under "Further Reading."

11.2 P–V Curves

Consider the two-bus power system of Figure 11.1. For simplicity, we model the line connecting these two buses as a pure inductive reactance jX, and assume that the generator maintains a constant voltage magnitude V_A. The complex power at the load is:

$$\overline{S} = P + jQ = \overline{V_B}\,\overline{I}^*$$

(11.1)

Power Systems: Fundamental Concepts and the Transition to Sustainability, First Edition. Daniel S. Kirschen.
© 2024 John Wiley & Sons Ltd. Published 2024 by John Wiley & Sons Ltd.
Companion website: www.wiley.com/go/kirschen/powersystems

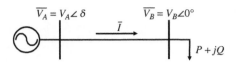

$$\overline{V_A} = V_A \angle \delta \qquad\qquad \overline{V_B} = V_B \angle 0°$$

Figure 11.1 A two-bus power system.

Taking $\overline{V_B}$ as the reference for the angles, we express the current $\overline{I}$ in terms of the voltages:

$$\overline{I} = \frac{\overline{V_A} - \overline{V_B}}{jX} = \frac{V_A \angle \delta - V_B \angle 0°}{jX} \tag{11.2}$$

Combining (11.1) and (11.2) we get:

$$\overline{S} = V_B \angle 0° \frac{V_A \angle -\delta - V_B \angle 0°}{-jX} = \frac{V_B V_A (\cos \delta - j \sin \delta) - V_B^2}{-jX}$$

$$= \frac{V_B V_A \sin \delta}{X} + j \frac{V_B V_A \cos \delta - V_B^2}{X} \tag{11.3}$$

Separating the real and imaginary parts of this expression we get:

$$P = \frac{V_B V_A}{X} \sin \delta \tag{11.4}$$

$$Q = \frac{V_B V_A \cos \delta - V_B^2}{X} \tag{11.5}$$

Using the trigonometric identity $\sin^2 \delta + \cos^2 \delta = 1$ to eliminate δ, we get the following quartic equation:

$$V_B^4 + \left(2QX - V_A^2\right) V_B^2 + \left(P^2 + Q^2\right)X^2 = 0 \tag{11.6}$$

Solving this equation for V_B, we have:

$$V_B = \sqrt{\frac{V_A^2}{2} - QX \pm \sqrt{\frac{V_A^4}{4} - QXV_A^2 - P^2X^2}} \tag{11.7}$$

Figure 11.2 shows how this voltage magnitude varies as a function of the active power load P while keeping the voltage magnitude V_A constant. Let us first consider this P–V curve for the unity power factor case. At no load, $V_B = V_A = 1.0$ pu. As the load increases, V_B decreases, at first slowly, then increasingly fast. For values of $P > 0.5$ pu, Eq. (11.7) no longer has a solution. This means that the system is fundamentally unable to supply a load greater than a critical value of 0.5 pu at unity power factor. The P–V curve for the 0.9 lagging power factor shows that for inductive loads the voltage decreases more rapidly as the load increases and that the maximum active power load that the system can supply is smaller. On the other hand, if the load has a leading power factor, the voltage initially rises as the load increases before decreasing as it approaches the maximum value that the system can deliver.

Example 11.1 *Critical loading* Consider the small system shown in Figure 11.3. Each line has an impedance of 0.6 pu and the generator maintains the voltage at bus A at its nominal value $V_A = 1.0$ pu. Calculate the maximum active power that this system can supply and the voltage at this critical load for load power factors of 1.0, 0.9 lag, and 0.9 lead.

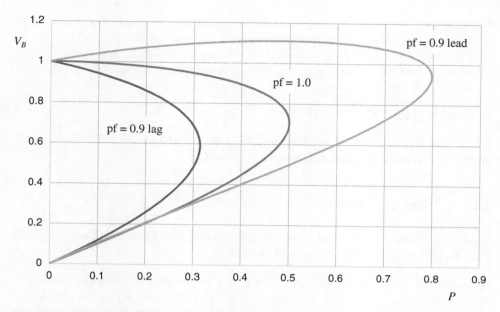

Figure 11.2 P–V curves for the system of Figure 11.1 with $V_A = 1.0$ pu and $X = 1.0$ pu.

Figure 11.3 System of Example 11.1.

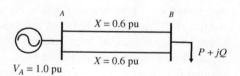

At the critical load the two solutions of (11.7) coincide. Therefore, at the critical load, we have: $\frac{V_A^4}{4} - QXV_A^2 - P^2X^2 = 0$. Inserting $X = \frac{0.6}{2} = 0.3$ pu; $V_A = 1.0$ pu, and $Q = \pm P \times \tan[\mathrm{acos(pf)}]$ in this expression, we get a quadratic equation. The positive solution of this equation is P^{crit}, which we insert in (11.7) to get V^{crit}, the voltage at the critical load.

Power factor	Q	p^{crit}	V^{crit}
Unity	0	1.6667	0.7071
0.9 lag	$0.4843 \times P$	1.0446	0.5901
0.9 lead	$-0.4843 \times P$	2.6591	0.9415

Normal system operation occurs along the upper part of these P–V curves, which corresponds to the "+" sign in (11.7). Operation along the lower part of the curve, (corresponding to the "−" sign in this expression) is undesirable as it is generally unstable and involves abnormally low voltages. Since voltage magnitudes below 0.95 pu are typically considered unacceptable, the practical limit on system loading is significantly smaller than the critical load.

Example 11.2 *Practical loading* For the unity power factor case in Example 11.1, we can easily calculate the practical maximum load, i.e., the load that would result in a 0.95 pu voltage at bus B. Setting $Q = 0$ in (11.7) and considering only the solution with the "+" sign as we want to operate on the upper part of the curve, we have:

$$V_B = \sqrt{\frac{V_A^2}{2} + \sqrt{\frac{V_A^4}{4} - P^2 X^2}}$$

To ensure that $V_B \geq 0.95$ pu, we must keep $P \leq 0.989$ pu, which is about 60% of the maximum load that the system can supply at unity power factor.

11.3 Effect of Outages

Sudden events, such as the outage of a transmission line or of a generating unit, can push a system beyond its stability limit, causing it to collapse.

Example 11.3 *Line disconnection* Figure 11.4 shows the P–V curves of the system of Example 11.1 when the load is at unity power factor with one or two lines in service. The black dot indicates the operating point of the system with two lines in service and $P = 0.989$ pu. As we calculated in Example 11.2, this loading would result in $V_B = 0.95$ pu and would be considered acceptable. However, if one of the lines were suddenly disconnected, the impedance X between the generation and the load would double and the P–V curve would shift to the left. The load would then exceed P^{crit} and the system would collapse. To avoid such a potentially disastrous outcome, the load should be limited to a value smaller than the $P^{crit} = 0.834$ pu for the system with two lines in service.

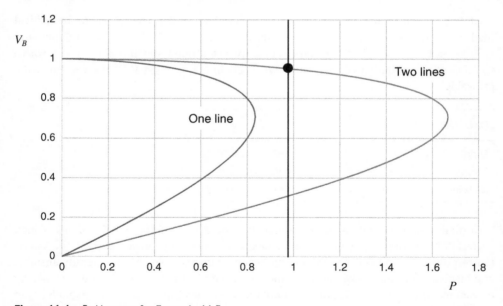

Figure 11.4 *P–V* curves for Example 11.3.

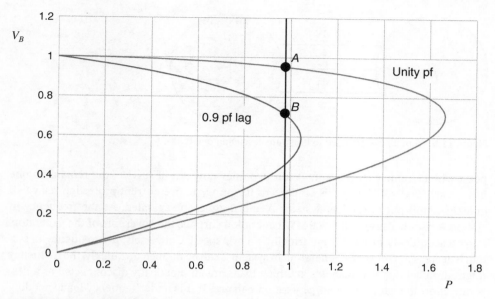

Figure 11.5 *P–V* curves for Example 11.4.

Example 11.4 *Loss of reactive compensation* Let us assume again that the system of Example 11.1 is operating at point A in Figure 11.5, i.e., unity power factor, $P = 0.977$ pu, two lines in service. Suppose that this unity power factor is achieved thanks to some reactive power compensation. If that reactive power compensation were to suddenly fail, the active power would remain unchanged, but the power factor would drop to 0.9 lag. The new operating point would then be at B, which is not only at an unacceptably low voltage but, more importantly, is perilously close to the critical load. A more severe loss of reactive compensation or a simultaneous increase in active load would cause a system collapse.

11.4 Cascading Overloads

Figure 11.6 shows two systems, or two parts of the same system, connected by three identical parallel lines. Each of these lines is designed to carry a maximum of 140 MVA. If system A exports 300 MW to system B, 100 MW flows in each line if we assume that the reactive power flows are negligible. Suppose that a fault occurs on one of these lines. The protection relays will quickly detect this fault and open the circuit breakers at both ends to de-energize this line. Since the generators in system A still produce 300 MW to be exported to system B, this power has to flow on the remaining two lines. Each of these lines now carries 150 MW, which exceeds their 140 MVA rating. Overhead transmission lines can sustain a small overload for some time. However, the increased current creates additional I^2R losses, which heat up the conductors and cause them to dilate and sag. Unless operators quickly reduce the flow on the two remaining lines by increasing generation in system B and decreasing it in system A to, this sag will continue to increase. After some time, the vertical distance between one of the conductors and whatever vegetation grows below will decrease to the

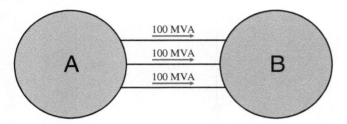

Figure 11.6 Simple system used to illustrate cascading overloads.

point where another fault occurs. With two lines now out of service, the remaining one has to carry 300 MW. Since this far exceeds its rating, an overcurrent protection relay will quickly take it out of service. Systems A and B are then separated. As the frequency in system A rises because of the 300 MW generation surplus, the governors of the generators must react quickly to avoid them tripping on overspeed. Conversely, the frequency in system B decreases. Unless the generation in system B can be ramped up quickly, the frequency might drop below the thresholds at which loads are automatically disconnected as a desperate effort to restore the load/generation balance. If this underfrequency load shedding fails to stop the decrease in frequency, generators would then trip because operation at low speed could damage the turbines. System B would then collapse.

Example 11.5 *Cascading overload* Consider the small system shown in Figure 11.7. Two of the lines have the same impedance and the same maximum rating of 50 MVA. The third line has twice the impedance and a maximum rating of 40 MVA. The load is 100 MVA at unity power factor. If the generator at bus *A* has a significantly lower operating cost than the generator at bus *B*, determine how much power each generator should produce to avoid the possibility of a cascading outage. Assume that the flows of reactive power are negligible.

If one of the lines rated at 50 MVA were suddenly disconnected, due to the difference in impedance of the two remaining lines, two-third of the power produced by generator at bus *A* would flow through the remaining 50 MVA line and one-third through the 40 MVA line. We must therefore ensure that: $\frac{2}{3}P_A \le 50$ or $P_A \le 75$ MW. To meet the load while avoiding the possibility of a cascading outage, the generator *B* must therefore produce 25 MW. Note that the 40 MVA limit is less constraining because $\frac{1}{3} \times 75 = 25 < 40$ MVA. Because it carries less power than the 50 MVA lines due to its larger impedance, the sudden disconnection of the 40 MVA line is not a concern.

Because the network is not meshed in these simple examples, we can easily identify the critical contingency and calculate the limitation it places on the operation of the system. In

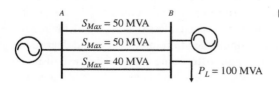

Figure 11.7 System of Example 11.5.

larger systems, a power flow calculation must be performed for each credible contingency to determine whether it would result in any flow exceeding the MVA rating of a line or transformer.

Case study – European incident of November 4, 2006

In the very northwest corner of Germany, on a river close to the border with the Netherlands, a shipyard builds cruise ships. To get to the North Sea, these ships must go under a double-circuit transmission line. However, the minimum distance between the conductors and the mast of the ship is small enough that the electrical insulation provided by air could break down. To avoid the risk of a fault, the line must therefore be de-energized while a ship sails pass the line. Prior to such a sailing scheduled for November 4, 2006, the system operators had performed power flow calculations to check that disconnecting this line would not cause overloads on other lines. However, these predictions turned out to be inaccurate and when the line was disconnected the operators noticed an overload in a line slightly to the South. They tried to remedy the problem by performing a switching operation. Unfortunately, this switching action made matters worse, and the overloaded transmission line tripped on overcurrent protection. At the time, there was a surplus of generation in northeastern Europe and this power flowed in a southwesterly direction toward Belgium, France, Spain, and Italy. The power that could no longer flow through the disconnected lines rerouted itself through other lines, causing them to trip on overcurrent protection. Within seconds, the continental European power system, which normally extends from Portugal to Poland, had split into three parts as shown in Figure 11.8.

As shown in Figure 11.9, the frequency in the Western part of the system dropped rapidly as it had a significant deficit of generation, while the frequency in the Northeastern part rose because of a surplus. Frequency in the Southeastern part dropped slightly because it had a more modest deficit. The inadvertent automatic disconnection of a substantial amount of wind generation exacerbated the generation imbalance in the Western part of the system. This imbalance was so severe that the frequency dropped below 49 Hz, which is the threshold that triggers automatic underfrequency load shedding. A total of about 18,600 MW of load was disconnected across eleven countries, including countries such as Portugal that are located far from the source of the incident. After this dramatic load shedding had stabilized the frequency, operators in the Western part of the system brought additional generation online to restore the frequency to its nominal value. Simultaneously, generators in the Northeastern part of the system reduced generation to lower its frequency and re-synchronize it with the Western part. This process took about 45 min. Re-synchronization with the Southeastern part of the system was accomplished shortly thereafter.

One of the main lessons learned from this incident is that operators must have access to good information not only about the part of the system for which they are responsible but also about how their actions might be affected by the situation in other parts of the system.

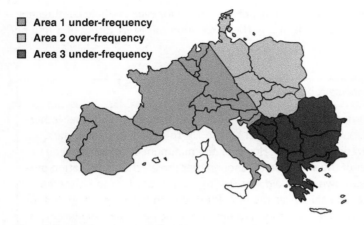

Area 1 under-frequency
Area 2 over-frequency
Area 3 under-frequency

Figure 11.8 Separation of the continental European power system as a result of the incident of November 4, 2006. Source: UCTE/https://eepublicdownloads.entsoe.eu/clean-documents/pre2015/publications/ce/otherreports/Final-Report-20070130.pdf.

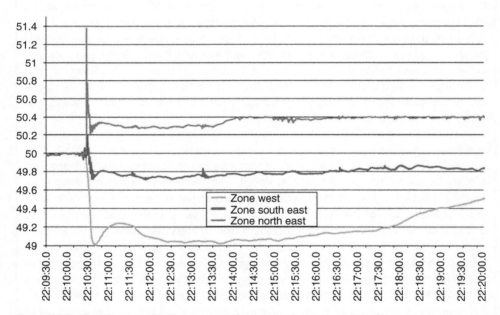

Figure 11.9 Frequency in the three areas following the separation of the European system. Source: UCTE/https://eepublicdownloads.entsoe.eu/clean-documents/pre2015/publications/ce/otherreports/Final-Report-20070130.pdf.

11.5 Electromechanical or Transient Stability

In Chapter 6, we argued that in the steady state the mechanical power provided to a generator by its prime mover is equal to the electrical power that this generator produces. This power balance ensures that the rotational speed of the generator remains constant. As long as changes in the mechanical power input and electrical power output of a generator are gradual, the generator remains in a quasi-steady state. However, an electrical fault close to the terminals of the generator can suddenly reduce its ability to inject active power in the network. Such an event creates a large imbalance between mechanical and electrical power. This imbalance pulls the generator away from its quasi-steady state and can create an instability in the system. To study the consequences of such faults, we must therefore consider the electromechanical dynamics of the generator as described by its differential equation of motion.

11.5.1 Modeling the Mechanical Dynamics of Synchronous Generators: the Swing Equation

As illustrated in Figure 11.10, a synchronous generator and its prime mover form a rotational mechanical system. T_m is the mechanical torque that the prime mover applies to the shaft connecting it to the generator minus the mechanical losses, while T_e is the electrical reaction torque that the generator develops in the opposite direction as it produces electrical power. Applying Newton's law of motion to this system, we have:

$$J\frac{d^2\theta_m}{dt^2} = T_m - T_e \tag{11.8}$$

where J is the moment of inertia of this mechanical system, and θ_m measures its angular position. When this system is in the steady state, it rotates at a constant speed, and we have:

$$\theta_{m,syn}(t) = \omega_{m,syn}\, t + \theta_{m,0} \tag{11.9}$$

where $\omega_{m,\,syn}$ is the synchronous speed, i.e., the constant speed at which the generator rotates when it produces a voltage at its rated frequency. However, in this section we are not interested in the steady-state behavior but rather on deviations from this steady state. Therefore, we introduce the angular deviation δ_m, which is the difference between the actual angular position of the mechanical system and the position it would have if it had rotated constantly at synchronous speed:

$$\delta_m(t) = \theta_m(t) - \theta_{m,syn}(t) \tag{11.10}$$

$\delta_m(t)$ can also be viewed as the angular position measured in a synchronously rotating reference frame. Taking the derivative of (11.10), we have:

$$\frac{d\delta_m}{dt} = \frac{d\theta_m}{dt} - \omega_{m,syn} = \omega_m(t) - \omega_{m,syn} \tag{11.11}$$

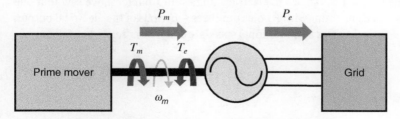

Figure 11.10 Illustration of the derivation of the swing equation.

where ω_m is the actual mechanical rotational speed. Since $\omega_{m,syn}$ is a constant if we take the derivative of (11.11), we get:

$$\frac{d^2\delta_m}{dt^2} = \frac{d^2\theta_m}{dt^2} \tag{11.12}$$

Using (11.12) we can rewrite (11.8) in terms of the angular deviation δ_m:

$$J\frac{d^2\delta_m}{dt^2} = T_m - T_e \tag{11.13}$$

To facilitate the integration of this model of the generator's mechanical behavior with a model of the electricity grid, we multiply both sides of this expression by ω_m to convert the torques to powers:

$$J\omega_m(t)\frac{d^2\delta_m}{dt^2} = \tilde{P}_m - \tilde{P}_e \tag{11.14}$$

where $\tilde{P}_m = \omega_m T_m$ is the mechanical power provided by the prime mover net of losses and $\tilde{P}_e = \omega_m T_e$ is the electrical power produced by the generator, including the electrical losses. We can then normalize this expression by dividing both side by the MVA rating of the generator S_{rated}:

$$\frac{J\omega_m(t)}{S_{rated}}\frac{d^2\delta_m}{dt^2} = P_m - P_e \tag{11.15}$$

where P_m and P_e are the mechanical and electrical powers expressed in per unit.

The moment of inertia J varies over a wide range depending on the rating of the generator and the type of prime mover. Since this can be inconvenient, manufacturers provide instead the inertia constant H, which is defined as the ratio of the kinetic energy stored in the system at synchronous speed over the volt-ampere rating of the generator:

$$H = \frac{\frac{1}{2}J\omega_{m,syn}^2}{S_{rated}} \tag{11.16}$$

H varies over a much smaller range than J, typically from 1 to 10 s. Replacing J in (11.15) by an expression extracted from (11.16), we get:

$$\frac{2H\omega_m(t)}{\omega_{m,syn}^2}\frac{d^2\delta_m}{dt^2} = P_m - P_e \tag{11.17}$$

To complete the integration of this mechanical equation with the electrical system, we need to convert the mechanical angles to electrical angles. In Chapter 4, we saw that one mechanical rotation of a generator with P poles produces $\frac{P}{2}$ periods of the electrical output. Electrical and mechanical angles and rotational speeds are therefore related as follows:

$$\omega_{syn} = \frac{P}{2}\omega_{m,syn} \tag{11.18}$$

$$\omega(t) = \frac{P}{2}\omega_m(t) \tag{11.19}$$

$$\delta(t) = \frac{P}{2}\delta_m(t) \tag{11.20}$$

Since the factor $\frac{2}{P}$ appears twice in the numerator and twice in the denominator on the left-hand side of (11.17), we can write rewrite this expression as follows:

$$\frac{2H\omega(t)}{\omega_{syn}^2}\frac{d^2\delta}{dt^2} = P_m - P_e \tag{11.21}$$

Finally, let us define the per unit speed as the ratio of the actual speed to the synchronous speed:

$$\omega_{pu}(t) = \frac{\omega(t)}{\omega_{syn}} \tag{11.22}$$

Inserting this expression in (11.21), we get the standard form of what is called the swing equation:

$$\frac{2H}{\omega_{syn}}\omega_{pu}(t)\frac{d^2\delta}{dt^2} = P_m - P_e \tag{11.23}$$

Using (11.11), we can transform this second-order differential equation into a system of two first-order differential equations:

$$\frac{d\delta}{dt} = \omega(t) - \omega_{syn} \tag{11.24}$$

$$\frac{d\omega}{dt} = \frac{\omega_{syn}}{2H\omega_{pu}(t)}(P_m - P_e) \tag{11.25}$$

This first-order formulation of the swing equation is more convenient when simulating the dynamic behavior of generators, particularly when using numerical integration.

When a generator and its prime mover are in the steady state, the mechanical power P_m and the electrical power P_e are equal. The right-hand side of (11.25) is then equal to zero, which means that the speed ω of the generator remains constant and equal to the synchronous speed ω_{syn}. The right-hand side of (11.24) is thus also equal to zero, which means that the angle δ remains constant. If a sudden and large decrease in the electrical power P_e makes the right-hand side of (11.25) positive, this equation shows that ω increases, i.e., the rotor accelerates. As ω becomes larger than the synchronous speed ω_{syn}, the angle δ also increases.

Using Eqs. (11.24) and (11.25) we can calculate how the speed and rotor angle of a generator change when, as illustrated in Figure 11.11, a bolted three-phase fault occurs at its terminals at $t = 0$. Such a fault forces the voltage at the terminal of the generator to zero and thus prevents it from injecting any electric power into the grid. Therefore, as long as the fault lasts, we have: $P_e = 0$ pu. On the other hand, we will assume that P_m remains unchanged. Solving these differential equations is much easier if we make the simplifying assumption that ω_{pu} remains constant at 1.0 pu. We will show that this is a reasonable approximation. We can thus rewrite (11.25) as follows:

$$\frac{d\omega}{dt} = P_m\frac{\omega_{syn}}{2H} \text{ with the initial condition } \omega(0) = \omega_{syn}$$

Integrating this equation gives:

$$\omega(t) = P_m\frac{\omega_{syn}}{2H}t + \omega_{syn}$$

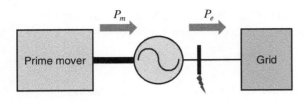

Figure 11.11 Bolted three-phase fault at the terminals of a generator.

Inserting this expression for $\omega(t)$ in (11.24), we have:

$$\frac{d\delta}{dt} = P_m \frac{\omega_{syn}}{2H} t \text{ where } \delta(0) \text{ is determined by the pre-fault loading of the generator.}$$

Integrating again, we get:

$$\delta(t) = P_m \frac{\omega_{syn}}{4H} t^2 + \delta(0) \tag{11.26}$$

Example 11.6 *Generator acceleration during a fault at its terminals* Using (11.26), we can calculate the time required for a 60 Hz generator with an inertia constant $H = 3$ s operating at rated power and a rotor angle $\delta(0) = 15° = 0.262$ rad to reach $\delta = 90°$ when a bolted three-phase fault occurs at its terminals.

With $P_m = 1.0$ pu and $\omega_{syn} = 2\pi f = 377.0$ rad/s, we have:

$$\omega(t) = 62.83 \, t + 377.0 \text{ rad/s}$$

and

$$\delta(t) = 31.42 \, t^2 + 0.262 \text{ rad}$$

The angle δ will thus reach 1.57 rad or 90° at $t = 204$ ms.

At that time, we have $\omega(t) = 389.82$ rad/s and $\omega_{pu} = \frac{\omega(t)}{\omega_{syn}} = 1.034$ pu, which justifies the simplifying assumption made earlier.

11.5.2 Modeling the Electrical Dynamics of Synchronous Generators

Because the electrical dynamics of synchronous generators are very complex, their detailed analysis is outside the scope of this book. We will therefore adopt a simplified model that is sufficient to describe the fundamentals of transient stability. As Figure 11.12 shows, the structure of this "classical" model is identical to the structure of the steady-state model of synchronous generators that we developed in Chapter 4 as it also consists of an ideal voltage source representing the internal emf, in series with a reactance. However, the values of the parameters X' and $\overline{E'}$ of this dynamic model differ from those of the steady-state model. In Chapter 9, we indeed argued that during transients the reactance of a synchronous machine varies because the path of the magnetic flux evolves during transients. It is convenient to consider three discrete values of this reactance as it changes over time: the subtransient reactance X'' immediately after the fault, followed by the transient reactance X', and then the synchronous reactance X_S for the steady state. Because the subtransient reactance X'' has the smallest value, it is the one to use when estimating the worst-case fault currents. On the other hand, electromechanical transients occur during the period where the transient

Figure 11.12 Classical dynamic model of a synchronous generator.

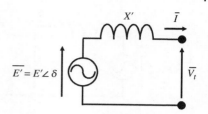

reactance X' is relevant. Our model therefore incorporates X'. If the generator is unloaded, $\overline{E'} = \overline{E}$. On the other hand, if the generator is producing power, $\overline{E'}$ must be adjusted to account for the demagnetizing effect of the stator current:

$$\overline{E'} = \overline{V_t} + jX'\overline{I} \tag{11.27}$$

As we discussed in Chapter 4, the internal emf $\overline{E'} = E'\angle\delta$ represents the voltage induced in the stator by the rotation of the magnetic field attached to the rotor. Its phase angle δ is thus tied to the physical position of the rotor. If we measure this position in a synchronously rotating reference frame (i.e., with respect to the stator voltage $\overline{V}$) this electrical phase angle δ is identical to the mechanical angle δ that we used in Section 11.5.1 to describe the position of the rotor. Our simplified dynamic electrical model of the synchronous generator is thus linked to its mechanical model (11.23) in two ways:

- Through the phase angle δ on the left-hand side
- Through the electrical power P_e on the right-hand side.

11.5.3 A Simple Model of the Rest of the System

Since stability is a characteristic of the entire system, the electromechanical models of all the generators must be integrated into a model of the entire system. We will describe how this integration is done in a Section 11.5.8 and how this system model is used to assess stability. Before doing that, it is useful to examine qualitatively how various factors influence stability. To this end, we will consider the simplest case, i.e., the interactions between a single generator and the rest of the system, where we assume that the rest of the system is perfectly stable. We model this stability by representing the rest of the system using an ideal voltage source, i.e., a source whose voltage magnitude, frequency, and phase angle are unaffected by any disturbance. This model is called "one machine versus infinite bus." As Figure 11.13 illustrates, in this model an inductive reactance X represents the transmission lines connecting the generator under study to the rest of the system.

11.5.4 Stable and Unstable Operating Points

As we showed in Section 4.8 of Chapter 4, the active electrical power supplied by the generator of Figure 11.13 to the rest of the system is given by the following expression:

$$P_e(\delta) = \frac{E'V_\infty}{X' + X}\sin\delta \tag{11.28}$$

In the steady state, if we neglect losses, the mechanical power P_m provided by the prime mover is equal to the electrical power P_e that the generator injects into the network. In

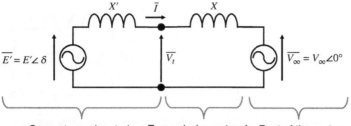

Figure 11.13 One machine versus infinite bus model.

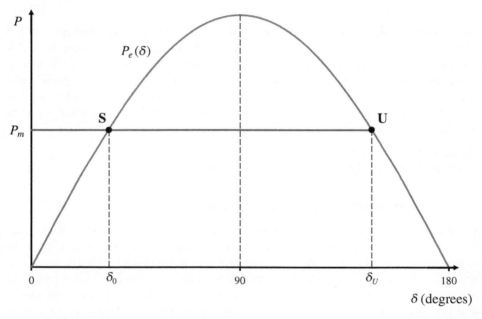

Figure 11.14 Stable and unstable equilibrium points of a synchronous generator connected to an infinite bus.

Figure 11.14, the intersections S and U of the curve representing (11.28) with the horizontal line at $P = P_m$ are thus equilibrium operating points. There is, however, a fundamental difference between these two equilibria: S is stable while U is unstable. To demonstrate this, Figure 11.15 illustrates how small disturbances affect operation around these operating points. Suppose first that the system is operating at point S and that a fluctuation in the network creates a small change $\Delta\delta$ around δ_0 while the mechanical power P_m remains constant. If $\Delta\delta > 0$, from an electrical perspective, the system is operating at point S_1. However, at that point the mechanical power is less than the electrical power and the generator slows down. Since δ represents both the position of the rotor and the phase angle, this slowdown reduces δ, counteracting the initial fluctuation and moving the operating point back toward the equilibrium point S. Conversely, if $\Delta\delta < 0$, the electrical operating point would be at S_2,

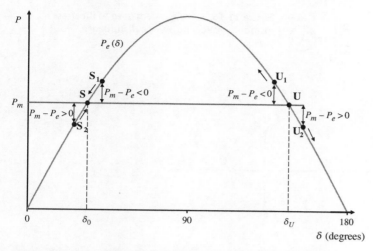

Figure 11.15 Stability analysis of the equilibrium operating points.

where P_m is greater than P_e. The generator would therefore accelerate, increase δ, and again move the operating point back toward S. Since small perturbations around S are resorbed, this equilibrium is deemed stable.

Consider now operation around the equilibrium point U. A small negative $\Delta\delta$ moves the electrical operating point to U_1 where P_e is greater than P_m. This imbalance slows the generator, reduces δ, and thus pulls U_1 further away from U toward S. If a positive $\Delta\delta$ moves the electrical operating point to U_2, the generator accelerates because the mechanical power is then larger than the electrical power. This acceleration further increases δ, pulling U_2 further away from U. Since any small perturbation around U gets amplified, it is an unstable equilibrium.

Example 11.7 *Stable and unstable operating points* A synchronous generator with a transient reactance $X' = 0.30$ pu is connected directly to an infinite bus. It delivers 1.0 pu apparent power at 0.95 pf lagging and nominal voltage. Calculate the rotor angles corresponding to the stable and unstable operating points.

$$\overline{S} = P + jQ = 1.0\angle\cos^{-1}(0.95) = \overline{V}\overline{I}^* \rightarrow \overline{I} = 1.0\angle - 18.195°\,\text{pu}$$

$$\overline{E'} = \overline{V}_t + jX'\overline{I} = 1.0\angle 0° + j0.30 \times 1.0\angle - 18.195° = 1.13\angle 14.6°\,\text{pu}$$

Hence $\delta_0 = 14.6°$.

(We can check this result using (11.28) with $X = 0$ and $P_m = P_e = 1.0 \times 0.95 = 0.95$ pu)

$$\delta_U = 180° - \delta_0 = 180 - 14.6 = 165.4°$$

11.5.5 Large Disturbances

In Section 11.5.4, we showed that small disturbances do not affect the stability of operating points where $\delta < 90°$. However, this stability is not absolute: a sufficiently severe disturbance will cause an instability. To study how this happens, consider the small system of

Figure 11.16 Small system used to illustrate instability following a large disturbance.

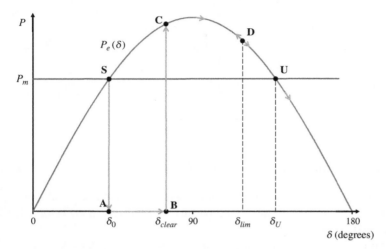

Figure 11.17 Stable and unstable evolutions of the system of Figure 11.16 following a fault.

Figure 11.16. Generator G is connected to the rest of the system (modeled as an infinite bus) by two lines. Circuit breaker CB_2 at the infinite bus end of one of these lines is open. Initially, generator A operates at point S of Figure 11.17. The angle δ_0 is such that:

$$P_m = \frac{E'V_\infty}{X'+X} \sin \delta_0 \tag{11.29}$$

where X is the reactance of the single line connecting generator G to the infinite bus.

Suppose that a three-phase bolted fault occurs on the open-ended line, just on the line side of circuit breaker CB_1. Such a fault forces the terminal voltage of the generator to zero. Because the generator cannot inject active power into a zero voltage, its electrical operating point moves instantly to point A in Figure 11.17. Since the mechanical power P_m is no longer matched by electrical power, the generator accelerates and the rotor angle δ increases. At time t_1, the protection system detects and clears the fault by opening circuit breaker CB_1. Since this restores the voltage, the generator can again inject power into the rest of the system. If by time t_1 the rotor angle had reached a value δ_{clear}, the electrical operating point jumps from B to C in Figure 11.17. At that point $P_e > P_m$ and the generator starts decelerating. However, it acquired momentum (kinetic energy) during the fault and its speed ω is larger than the synchronous speed ω_{syn}. The angle δ will thus continue to increase. As long as δ is such that $P_e > P_m$, this extra kinetic energy gets transformed into electrical energy and the rotor decelerates. Depending on the initial conditions and the duration of the fault, the system will either stabilize or go unstable:

- If as shown in Figure 11.17, $\delta = \delta_{lim} < \delta_U$ when the extra kinetic energy has been fully injected into the rest of the system, ω then drops below ω_{syn}, δ starts to decreases, and the system returns toward the stable operating point S.
- On the other hand, if δ increases beyond δ_U, P_e drops again below P_m. The rotor then reaccelerates rapidly. Protection devices detect this unstable condition and disconnect the generator to prevent damage.

Stability thus depends on whether the generator is able to rapidly inject into the grid the kinetic energy that it acquires during the fault. In Section 11.5.6, we develop a simple stability criterion that will help us visualize the factors that influence stability.

11.5.6 Equal Area Criterion

To develop this criterion, we start by assuming that $\omega_{pu}(t) \sim 1.0$ in (11.25):

$$\frac{2H}{\omega_{syn}} \frac{d\omega}{dt} = P_m - P_e \tag{11.30}$$

We then integrate both sides of this equation between two arbitrary angles δ:

$$\frac{2H}{\omega_{syn}} \int_{\delta_a}^{\delta_b} \frac{d\omega}{dt} d\delta = \int_{\delta_a}^{\delta_b} (P_m - P_e) d\delta \tag{11.31}$$

On the left-hand side of (11.31), we can swap $d\delta$ and $d\omega$ and integrate over ω instead of δ:

$$\frac{2H}{\omega_{syn}} \int_{\omega_a}^{\omega_b} \frac{d\delta}{dt} d\omega = \int_{\delta_a}^{\delta_b} (P_m - P_e) d\delta \tag{11.32}$$

where ω_a and ω_b are the speeds of the generator when its rotor angle is, respectively, δ_a and δ_b. Since $d\delta/dt = \omega$, we have:

$$\frac{2H}{\omega_{syn}} \int_{\omega_a}^{\omega_b} \omega \, d\omega = \int_{\delta_a}^{\delta_b} (P_m - P_e) d\delta \tag{11.33}$$

Calculating the integral on the left-hand side, we get:

$$\frac{H}{\omega_{syn}} \left(\omega_b^2 - \omega_a^2 \right) = \int_{\delta_a}^{\delta_b} (P_m - P_e) d\delta \tag{11.34}$$

Since the factor $\frac{H}{\omega_{syn}}$ is proportional to the moment of inertia of the generator, the left-hand side is proportional to the change in kinetic energy as the rotor angle varies between δ_a and δ_b. The right-hand side of (11.34) shows that this change in kinetic energy is proportional to the area between P_m and the P_e curve. In Figure 11.18, area A_1 is thus proportional to the kinetic energy that the generator acquires during the fault, while area A_2 represents the kinetic energy that the generator can return to the system before going unstable. The system will therefore remain stable if $A_1 \leq A_2$. As Figure 11.19 illustrates, there is thus a critical clearing angle δ_{crit} such that:

$$A_1 = A_2 \tag{11.35}$$

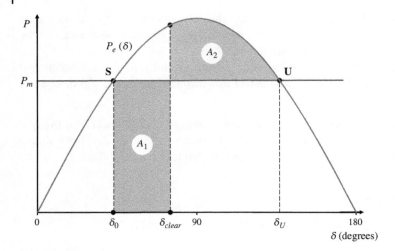

Figure 11.18 Increase and decrease in kinetic energy following a fault.

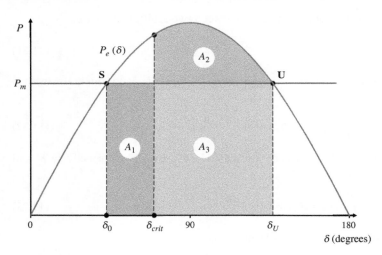

Figure 11.19 Equal area criterion.

To facilitate the calculation of δ_{crit}, we can rewrite this expression as follows:

$$A_1 + A_3 = A_2 + A_3 \tag{11.36}$$

Using the notations of Figure 11.19, we can express this equal area criterion in terms of the angles and the powers:

$$P_m \times (\delta_U - \delta_0) = \int_{\delta_{crit}}^{\delta_U} P_e(\delta)\, d\delta \tag{11.37}$$

Replacing $P_e(\delta)$ by its expression (11.28), we have:

$$P_m \times (\delta_U - \delta_0) = \frac{EV_\infty}{X' + X} \int_{\delta_{crit}}^{\delta_U} \sin\delta\, d\delta \tag{11.38}$$

Calculating the integral to derive an expression for δ_{crit}:

$$P_m \times (\delta_U - \delta_0) = \frac{EV_\infty}{X' + X}(-\cos \delta_U + \cos \delta_{crit}) \tag{11.39}$$

$$\delta_{crit} = \cos^{-1}\left(\frac{X' + X}{EV_\infty}P_m \times (\delta_U - \delta_0) + \cos \delta_U\right) \tag{11.40}$$

Example 11.8 *Equal area criterion* The 60 Hz generator of Figure 11.16 has a transient reactance $X' = 0.3$ pu and an inertia constant $H = 3$ s. It is connected to the rest of the system, modeled as an infinite bus, by two lines modeled by an inductive reactance $X = 0.4$ pu. One of these lines is disconnected at the infinite bus. The voltage magnitude at the infinite bus is at nominal value, while the magnitude of the transient internal emf of the generator $E' = 1.2$ pu This generator injects 1.0 pu of active power into the rest of the system. Let us assess the stability of this generator if a bolted fault occurs as shown on this figure.

Before the fault, we have: $P_m = P_e = \frac{E'V_\infty}{X'+X} \sin \delta_0$. Hence $\delta_0 = \sin^{-1}\left(\frac{1.0 \times (0.3+0.4)}{1.2 \times 1.0}\right) = 35.69°$.
From Figure 11.14 we observe that $\delta_U = 180° - \delta_0 = 144.31°$.
Applying (11.40) taking care of converting degrees into radians, we have:

$$\delta_{crit} = \cos^{-1}\left(\frac{0.3 + 0.4}{1.2 \times 1.0} \times 1.0 \times [144.31° - 35.69°] \times \frac{\pi}{180} + \cos 144.31°\right) = 72.92°$$

In Example 11.6 we showed that during such a fault, the rotor angle of the generator increases according to the following expression:

$$\delta(t) = P_m \frac{\omega_{syn}}{4H} t^2 + \delta(0)$$

Given $P_m = 1.0$ pu; $\omega_{syn} = 377.0$ rad/s; $H = 3$ s; and $\delta(0) = \delta_0 = 35.69°$, $\delta_{crit} = 72.92°$, it would take 144 ms to reach the critical clearing angle. This means that the fault would have to be cleared in less than 144 ms for the system to avoid instability.

For simplicity, we have assumed so far that, once the fault has been cleared, the system returns to its pre-fault condition. Since this is usually not the case, it is important to distinguish the pre-fault, fault, and post-fault conditions. Figure 11.20 illustrates these conditions for a one-machine infinite bus system. Before the fault, the generator injects power into the rest of the system through two parallel lines of equal reactance X. During the fault, it cannot supply any electric power because the voltage at its terminals is zero. After the fault is cleared by disconnecting the faulted line, the generator can again inject power in the system, albeit through a single line of reactance X. Before the fault, the total reactance between the internal emf of the generator and the infinite bus is thus $X' + X/2$, while it is $X' + X$ after the fault has been cleared. As Figure 11.21 illustrates, the post-fault power angle curve is therefore below the pre-fault curve. On this figure, the pre-fault operating point is at A. The fault moves it to B and on to C as the rotor angle increases. Clearing the fault puts the operating point at D on the post-fault curve. The extra kinetic energy acquired by the generator during the fault pushes the operating point up to E, where the angle starts decreasing because the electrical power is greater than the mechanical power. Finally, the system settles at the new stable operating point F. In this example, the angle δ_{clear} is such that the areas A_1 and A_2 are equal for an angle $\delta_{lim} < \delta_U$, which ensures that the system remains stable.

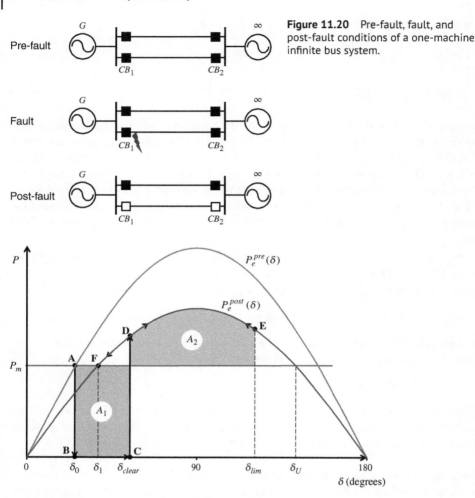

Figure 11.20 Pre-fault, fault, and post-fault conditions of a one-machine infinite bus system.

Figure 11.21 Pre-fault and post-fault power angle curves for the system of Figure 11.20.

Example 11.9 *Equal area criterion* As in Example 11.8, the 60 Hz generator of Figure 11.20 has a transient reactance $X' = 0.3$ pu and an inertia constant $H = 3$ s. The voltage magnitude at the infinite bus is at nominal value, while the magnitude of the internal emf of the generator $E' = 1.2$ pu. This generator injects 1.0 pu of active power into the rest of the system. Each of the two lines has a reactance $X = 0.4$ pu.

The pre-fault power angle curve is $P_e^{pre}(\delta) = \frac{E'V_\infty}{X' + \frac{X}{2}} \sin \delta = 2.4 \sin \delta$. At the pre-fault equilibrium, $P_m = 1.0 = P_e^{pre} = 2.4 \sin \delta_0$. Hence $\delta_0 = 24.62°$.

The post-fault power angle curve is: $P_e^{post}(\delta) = \frac{E'V_\infty}{X' + X} \sin \delta = 1.71 \sin \delta$. At the post-fault equilibrium, $P_m = P_e^{post} = 1.71 \sin \delta_1 = 1.0$. Hence $\delta_1 = 35.69°$ and $\delta_U = 180° - \delta_1 = 144.31°$.

To calculate the critical clearing angle, we apply (11.40), taking care of converting degrees into radians:

$$\delta_{crit} = \cos^{-1}\left(\frac{0.3 + 0.4}{1.2 \times 1.0} \times 1.0 \times [144.31° - 24.62°] \times \frac{\pi}{180} + \cos 144.31°\right) = 66.02°$$

We can then use (11.26) to calculate the critical time $t_{crit} = 152$ ms. This time is larger than in Example 11.8 because the generator starts from a smaller δ_0.

Example 11.10 *Equal area criterion* Suppose that in the system of Example 11.9, the three-phase bolted fault occurs half-way down one of the lines instead of close to the generator terminals, as illustrated in the equivalent circuit diagram shown at the top of Figure 11.22. While such a fault would cause the voltage at the terminals to sag, it would not force it down to zero. The generator would therefore continue to inject power in the rest of the system during the fault. To quantify this power transfer, we begin by replacing the portion of the equivalent circuit surrounded by dashes by its Thevenin equivalent:

$$X_{TH} = 0.2 \parallel 0.4 = 0.133 \text{ pu}$$

$$V_{TH} = \frac{j0.2}{j0.2 + j0.4} \times 1.0 = 0.333 \text{ pu}$$

The power transferred during the fault is then given by the following expression:

$$P_e^{fault}(\delta) = \frac{E' V_{TH}}{X' + X_{TH}} \sin \delta = 0.923 \sin \delta$$

As calculated in Example 11.9, $P_e^{post}(\delta) = 1.71 \sin \delta$

Figure 11.23 illustrates the evolution of the system during the fault and after the fault has been cleared with $\delta_{clear} < \delta_{crit}$. Because the generator is able to inject power into the rest of the system during the fault, the amount of kinetic energy that it acquires (area A_1) is much smaller than in the previous examples. To calculate δ_{crit}, we express the equality of areas A_1 and A_2 for $\delta_{lim} = \delta_U$:

$$\int_{\delta_0}^{\delta_{crit}} \left[P_m - P_e^{fault}(\delta) \right] d\delta = \int_{\delta_{crit}}^{\delta_U} \left[P_e^{post}(\delta) - P_m \right] d\delta$$

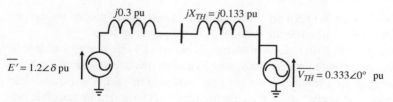

Figure 11.22 Impedance diagram of Example 11.10 and Thevenin equivalent of the faulted portion of the system.

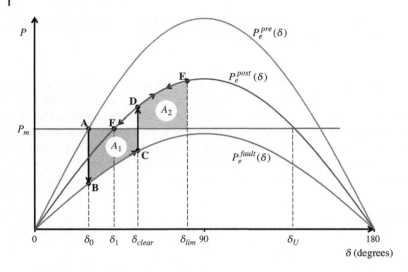

Figure 11.23 Pre-fault, fault, and post-fault power angle curves for the system of Example 11.10.

Which we can rewrite as follows:

$$\int_{\delta_0}^{\delta_{crit}} P_m d\delta + \int_{\delta_{crit}}^{\delta_U} P_m d\delta = \int_{\delta_0}^{\delta_{crit}} P_e^{fault}(\delta)d\delta + \int_{\delta_{crit}}^{\delta_U} P_e^{post}(\delta)d\delta$$

$$\int_{\delta_0}^{\delta_U} P_m d\delta = \int_{\delta_0}^{\delta_{crit}} 0.923 \sin \delta d\delta + \int_{\delta_{crit}}^{\delta_U} 1.71 \sin \delta d\delta$$

Calculating these integrals gives:

$$P_m(\delta_U - \delta_0) = 0.923(\cos \delta_0 - \cos \delta_{crit}) + 1.71(\cos \delta_{crit} - \cos \delta_U)$$

Inserting in this expression the values P_m, δ_U, and δ_0 from Example 11.9 gives $\delta_{crit} = 100.09°$, which shows that the rotor angle can temporarily exceed 90° without causing an instability. This critical clearing angle is significantly larger than when the fault occurs at the terminals of the generator. Note that we cannot use (11.26) to calculate the critical clearing time because the active power injected by the generator is not zero during the fault.

11.5.7 Factors Influencing Stability

While the examples of Section 11.5.6 are based on a simplistic system model, we can use them to discuss factors that influence the stability of actual power systems.

First, a fast clearing of the fault reduces the amount of extra kinetic energy acquired by the generator. It also gives more time for the system to absorb this energy before it reaches an unstable operating point. Promptly disconnecting faulted components is thus essential to maintaining stability. However, since the clearing time is determined by how fast protection relays can detect a fault and how fast circuit breakers can disconnect the affected component, it is set by the design of these components.

Second, an increase in the amount of power produced by a generator increases the risk of instability, not only because this power amplifies the kinetic energy that it acquires during the fault, but also because it increases the initial angle δ_0 and thus decreases the stability margin. In some systems, operators must therefore occasionally limit the power output of some generators to avoid a potential instability.

Third, generators connected to the rest of the system through long lines are more prone to instabilities. Long lines indeed have a higher reactance X, which increases the initial angle δ_0 and reduces the generator's ability to offload surplus kinetic energy. To improve stability, remote generators are connected to the rest of the system through extra-high-voltage (EHV) lines because such lines have a lower per unit reactance. Inserting capacitors in series with a line compensates for its inductance and reduces its apparent reactance. Providing parallel paths also reduces the overall reactance before and after the fault. When building multiple ac lines is not feasible or economical, HVDC lines provide an alternative way of transmitting large amounts of power from remote location without creating a risk of instability.

Finally, a fault at or near the terminals of a generator is more likely to affect stability than a fault further out in the system. Since we obviously have no control over where faults occur, we have to assume the worst-case scenario when assessing stability.

11.5.8 Transient Stability Analysis Using Time Domain Simulation

While the equal area criterion provides valuable insights, it is only applicable to very simple cases. A more systematic approach involves solving the differential Eqs. (11.24) and (11.25) to determine whether the rotor angle δ increases uncontrollably or returns to a stable value after a fault. Because these equations are nonlinear, they do not have an analytical solution and must be integrated numerically. Before applying numerical integration to these equations, let us review the simplest, albeit not the most accurate, technique: Euler's method.

Equation (11.41) shows the general form of a first-order differential equation.

$$\frac{dx(t)}{dt} = f(x); \quad x(0) = x_0 \tag{11.41}$$

Since we are given the initial condition $x(0) = x_0$, we can calculate the derivative of the function $x(t)$ at $t = 0$. This derivative is the slope of the tangent to the unknown function $x(t)$ at that point, as illustrated in Figure 11.24. For a sufficiently small timestep Δt, we can approximate the function $x(t)$ by its tangent and write:

$$x(\Delta t) \sim x_0 + f(x_0) \times \Delta t = x_1$$

Assuming that x_1 is a sufficiently good approximation of $x(\Delta t)$, we can repeat the process:

$$x_2 = x_1 + f(x_1) \times \Delta t$$

More generally:

$$x_{k+1} = x_k + f(x_k) \times \Delta t \tag{11.42}$$

Using Eq. (11.42) and the initial value x_0, we can calculate a series of approximate values x_k of the function $x(t)$ at discrete intervals Δt.

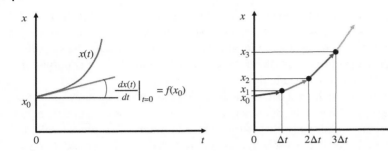

Figure 11.24   Illustration of the principle of Euler's method of numerical integration.

Example 11.11   *Euler's method*   The table below shows the first 10 steps in the numerical integration of the following first-order differential equation using Euler's method:

$$\frac{dx(t)}{dt} = 2x; \quad x(0) = 1.0$$

The time step is $\Delta t = 0.01$. For comparison, the last column of the table shows the values calculated using the analytical solution of this equation: $x(t) = e^{2t}$.

k	t	$f(x_k) = 2x_k$	x_k	$x(t) = e^{2t}$
0	0	2.0	1.0	1.0
1	0.01	2.04	1.02	1.02020134
2	0.02	2.0808	1.0404	1.04081077
3	0.03	2.122416	1.061208	1.06183655
4	0.04	2.16486432	1.08243216	1.08328707
5	0.05	2.20816161	1.10408080	1.10517092
6	0.06	2.25232484	1.12616242	1.12749685
7	0.07	2.29737134	1.14868567	1.15027380
8	0.08	2.34331876	1.17165938	1.17351087
9	0.09	2.39018514	1.19509257	1.19721736
10	0.1	2.43798884	1.21899442	1.22140276

Comparing the results of the numerical integration with the analytical solution shows that the error progressively increases. To reduce this error our only recourse when using Euler's method is to adopt a smaller step size. For example, if we use $\Delta t = 0.001$ instead of $\Delta t = 0.01$, at $t = 0.1$, we get $x_k = 1.22115883$. However, getting there required a hundred steps instead of 10 and thus increased the computing burden. More sophisticated numerical integration methods achieve better accuracy than Euler's method with a larger timestep.

Let us now consider how we can apply numerical integration to the study of transient stability. For convenience, the first-order formulation (11.24) and (11.25) of the swing equation is copied here:

$$\frac{d\delta}{dt} = \omega(t) - \omega_{syn} \tag{11.43}$$

$$\frac{d\omega}{dt} = \frac{\omega_{syn}}{2H\omega_{pu}(t)}[P_m - P_e(\delta)] \tag{11.44}$$

Observe first that these two equations are coupled: ω appears in the computation of the derivative of δ and δ affects the calculation of the derivative of ω through the electrical power P_e. These equations must therefore be integrated simultaneously, which means that at each step both equations must incorporate the value calculated by the other at the previous step. Applying Euler's integration formula (11.42), we have:

$$\delta_{k+1} = \delta_k + (\omega_k - \omega_{syn}) \times \Delta t \tag{11.45}$$

$$\omega_{k+1} = \omega_k + \frac{\omega_{syn}^2}{2H\omega_k}[P_m - P_e(\delta_k)] \times \Delta t \tag{11.46}$$

With the initial conditions:

$$\omega_0 = \omega_{syn} \tag{11.47}$$

$$P_e^{pre} = \frac{EV}{X_S}\sin\delta_0 = P_m \tag{11.48}$$

When integrating these equations to assess transient stability through simulation, we must distinguish three phases:

- *Pre-fault conditions:* The system is assumed to be in the steady state with $P_e^{pre} = P_m$. Since the second term of the right-hand side of (11.46) is equal to zero, the speed ω does not change and remains equal to the synchronous speed ω_{syn}. The right-hand side of (11.45) is therefore also equal to zero, which means that the angle δ remains equal to δ_0.
- *Fault conditions:* The electrical power P_e drops to $P_e^{fault} < P_m$. The second term of the right-hand side of (11.46) is now positive, which drives an increase in speed. Since the speed is now larger than the synchronous speed, the right-hand side of (11.45) is positive and the angle δ increases.
- *Post-fault conditions:* The electrical power P_e jumps to $P_e^{post} > P_m$. The second term of the right-hand side of (11.46) becomes negative. The speed starts to decrease, but the angle continues to increase until ω becomes less than ω_{syn}. If this happens before δ reaches δ_U, the system remains stable. Otherwise, the power balance reverses once more, and the speed and angle again start increasing.

Example 11.12 *Effect of the fault clearing time* Figure 11.25 illustrates three time-domain simulations of the system of Example 11.10 using Euler's method with a time step $\Delta t = 0.001$ s. Until the onset of the fault at $t = 0.1$ s, the system is in the steady state and the angle δ remains constant. In the first case, the fault is cleared at $t = 0.25$ s and the system remains stable. In the second case, the fault is cleared at $t = 0.411$ s, δ gets very close to the value of δ_U that we calculated in Example 11.9 but never exceeds it before ultimately decreasing. In this case, the system is marginally stable. In the third case, the fault is cleared at $t = 0.412$ s, δ exceeds δ_U at about $t = 0.615$ s, the rotor then re-accelerates, and the system goes unstable. The stable and marginally stable cases display

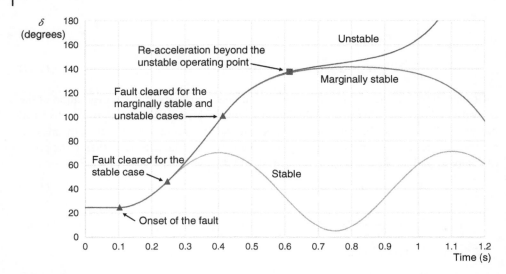

Figure 11.25 Time-domain simulation of the system of Example 11.10 for three different fault clearing times.

undamped oscillations after the fault is cleared because we have not included in our model the damping that would exist in an actual system.

Example 11.13 *Effect of the generator loading* Figure 11.26 shows three additional simulations of the system of Example 11.10. In each case, the fault occurs at $t = 0.1$ s and is cleared at $t = 0.25$ s. The stable case is identical to the stable case of Example 11.12 with a pre-fault generator power output $P_e^{pre} = 1.0$ pu. The second simulation shows that the system is marginally stable for $P_e^{pre} = 1.318$ pu as the system starts from a larger initial angle $\delta_0 = 33.31°$. In the third simulation, the pre-fault generator output is increased to $P_e^{pre} = 1.32$ pu, which causes the generator to go unstable.

11.5.9 Simulating the Dynamics of Multi-generator Systems

So far, our discussion of transient stability has focused on a single generator and assumed that the rest of the system was unaffected by the fault. In practice, a fault impacts all generators and alters to different degrees the amount of power that each of them injects into the system. These changes in injections affect the flows in the network and hence the voltage magnitudes and angles at all buses. The transients in each of the generators thus interact with the transients in the others through the network. To accurately assess the transient stability of a power system, we must take these interactions into account.

Figure 11.27 illustrates a multi-generator system where each generator is modeled as a constant voltage E' behind its transient reactance X'. The speed and rotor angle of each of these generators are governed by its swing equation. If there are n generators in the system, we thus have to solve $2n$ first-order differential equations of the form:

$$\frac{d\delta_i}{dt} = \omega_i(t) - \omega_{syn} \tag{11.49}$$

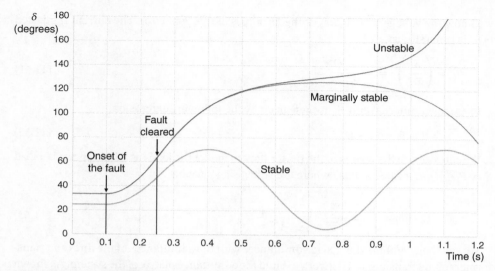

Figure 11.26 Time-domain simulation of the system of Example 11.10 for three different pre-fault generator output powers.

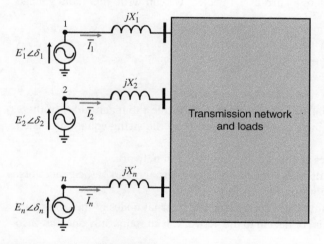

Figure 11.27 Schematic representation of a multi-machine system.

$$\frac{d\omega_i}{dt} = \frac{\omega_{syn}}{2H_i\omega_{i,pu}(t)}[P_{m,i} - P_{e,i}(t)] \tag{11.50}$$

where the electrical power $P_{e,i}$ injected by each generator into the network is a function not only of its own rotor angle δ_i but also of the state of the entire system at each instant. Before deriving an expression for $P_{e,i}$, we must determine the initial conditions of these equations and construct the network equations.

To determine the initial conditions, we perform a power flow calculation based on the pre-fault configuration of the network. Suppose that generator i is connected to bus k and injects the complex power $\overline{S}_i = P_i + jQ_i$ in the network. From the results of the power flow, we know the voltage $\overline{V}_k = V_k \angle \delta_k$ at the terminals of this generator. The current $\overline{I}_i$ injected by this generator is:

$$\overline{I}_i = \left(\frac{\overline{S}_i}{\overline{V}_k} \right)^* = \frac{P_i - jQ_i}{V_k \angle - \delta_k} \tag{11.51}$$

Its transient internal emf $\overline{E}'_i$ is then given by the following expression:

$$\overline{E}'_i = E'_i \angle \delta_i = \overline{V}_k + jX'_i \overline{I}_i \tag{11.52}$$

We can also use the results of the power flow to model the loads as admittances. If a load $\overline{S}_l = P_l + jQ_l$ is located at bus l where the voltage is $\overline{V}_l$, we have:

$$\overline{y}_l = g_l + jb_l = \frac{\overline{I}_l}{\overline{V}_l} = \frac{\left(\frac{\overline{S}_l}{\overline{V}_l} \right)^*}{\overline{V}_l} = \frac{P_l - jQ_l}{V_l^2} \tag{11.53}$$

We can then combine these load admittances with the admittances of the lines and transformers of the transmission network to build the admittance matrix of the system. As shown in Figure 11.27, the internal nodes of the generators are numbered 1 to n. The buses of the transmission network are given node numbers from $n+1$ to $n+1+m$. Since the generator nodes are the only nodes where a current is injected, we can write the nodal voltage equations as follows:

$$\begin{pmatrix} I \\ 0 \end{pmatrix} = \begin{pmatrix} Y_{nn} & Y_{nm} \\ Y_{mn} & Y_{mm} \end{pmatrix} \begin{pmatrix} E \\ V \end{pmatrix} \tag{11.54}$$

where:

I is the $n \times 1$ vector of currents injected by the generators.

E is the $n \times 1$ vector of internal emf of the generators. While the magnitudes of these voltages are assumed constant, their angles are governed by the swing equation of each generator.

V is the $m \times 1$ vector of voltages at buses in the transmission network.

Y_{nn} is the $n \times n$ matrix of admittances connected to the internal nodes of the generators. This matrix is diagonal, and its diagonal terms are $1/jX'_i$.

Y_{nm} is the $n \times m$ matrix of admittances connecting the internal nodes of the generators to the bus where each generator is connected to the network. It contains only one non-zero $-1/jX'_i$ term per row.

Y_{mn} is an $m \times n$ matrix that is the transpose of Y_{nm}.

Y_{mm} is the $m \times m$ admittance matrix of the transmission network, augmented by the equivalent admittances of the loads, as calculated using (11.53).

Once we have built the admittance matrix of the pre-fault system, we can modify it to reflect the faulted and post-fault conditions. Let us assume a worst-case scenario: a solid fault close to node i on line $i - j$. During the fault, the voltage at node i is zero, which means that this node merges with the reference or ground node. Therefore, we model this faulted condition by removing row i and column i from the pre-fault admittance matrix. Note that the admittances of all the lines connected to node i still appear in this matrix as part of the diagonal terms corresponding to the nodes at the other end of these lines. Since clearing the fault involves disconnecting line $i - j$, we model the post-fault conditions by removing the admittance of this line from the diagonal and off-diagonal terms of the pre-fault admittance matrix.

Since we are interested in how the generators interact with each other following a fault, we can reduce the $(n+m) \times (n+m)$ set of equations (11.54) to an $n \times n$ set that involves only the internal nodes of the generator. To do this, we expand (11.54) as follows:

$$I = Y_{nn}E + Y_{nm}V \tag{11.55}$$

$$0 = Y_{mn}E + Y_{mm}V \tag{11.56}$$

Since Y_{mm} is the admittance matrix of a physical network, it is invertible, and we can use (11.56) to express V as a function of E:

$$V = -Y_{mm}^{-1}Y_{mn}E \tag{11.57}$$

Inserting (11.57) into (11.55), we get:

$$I = \left(Y_{nn} - Y_{nm}Y_{mm}^{-1}Y_{mn} \right) E = \tilde{Y}E \tag{11.58}$$

Note that we have separate versions of the reduced admittance matrix $\tilde{Y}$: one for the pre-fault conditions, one for the faulted conditions, and one for the post-fault conditions. Row i of (11.58) expresses the current injected by generator i as a function of the internal emfs of all the generators in the system:

$$\overline{I}_i = \sum_{j=1}^{n} \tilde{Y}_{ij}\overline{E}_j = \sum_{j=1}^{n} (\tilde{G}_{ij} + j\tilde{B}_{ij})E_j\angle\delta_j \tag{11.59}$$

The active power injected by generator i is thus given by:

$$P_{e,i} = \text{Re}(\overline{S}_i) = \text{Re}\left(\overline{E}_i\overline{I}_i^* \right) = \text{Re}\left(E_i\angle\delta_i \sum_{j=1}^{n} (\tilde{G}_{ij} - j\tilde{B}_{ij})E_j\angle - \delta_j \right) \tag{11.60}$$

Expanding (11.60) gives us an expression for $P_{e,i}$ that we can use to calculate the right-hand side of (11.50):

$$P_{e,i} = E_i^2\tilde{G}_{ii} + \sum_{\substack{j=1 \\ j \neq i}}^{n} E_iE_j[\tilde{B}_{ij}\sin(\delta_i - \delta_j) + \tilde{G}_{ij}\cos(\delta_i - \delta_j)] \tag{11.61}$$

We can now outline a procedure to simulate the transient behavior of a multi-generator system.

Data requirements

- Parameters of all the generators: H_i, X_i', pre-fault voltage at the generators' terminals
- Parameters of all the lines and transformers
- Active and reactive load at each node: P_i, Q_i
- Location of the fault
- Time of fault onset t_{fault}, time of fault clearing t_{clear}, and time of end of simulation t_{end}
- Time step of the simulation Δt.

Initialization

1. Perform a power flow calculation assuming pre-fault conditions.
2. Using the results of this power flow, set the mechanical power of each generator $P_{m,i}$ equal to its electrical power output $P_{e,i}$
3. Using (11.51) and (11.52) and the results of this power flow, calculate the E_i and the initial value of δ_i for all generators.
4. Using (11.53) and the results of this power flow, calculate the equivalent admittance of each load.
5. Build the admittance matrix for the pre-fault conditions according to (11.54).
6. Modify the pre-fault admittance matrix to obtain the admittance matrix for the faulted condition.
7. Modify the pre-fault admittance matrix to obtain the admittance matrix for the post-fault condition.
8. Using (11.58), calculate the reduced admittance matrices $\tilde{Y}$ for the pre-fault, faulted, and post-fault conditions.
9. Set $t = 0$
10. For all generators, set $\omega_i(0) = \omega_{syn}$.
11. For all generators, set $\delta_i(0)$ to the value calculate in step 3.

Step-by-step simulation

1. Set $t = t + \Delta t$
2. If $t > t_{end}$, exit
3. Calculate $P_{e,i}(t)$ for each generator using the latest values of $\delta_j(t), j = 1 \cdots n$
   a. If $t < t_{fault}$, $P_{e,i}(t) = P_{m,i}$.
   b. If $t_{fault} \leq t < t_{clear}$, $P_{e,i}(t)$ is calculated using in (11.61) the elements of the reduced admittance matrix for the faulted conditions.
   c. If $t_{clear} \leq t < t_{end}$, $P_{e,i}(t)$ is calculated using in (11.61) the elements of the reduced admittance matrix for the post-fault conditions.

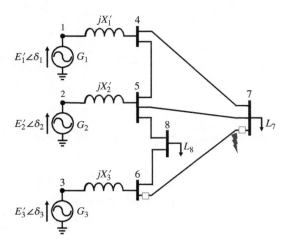

Figure 11.28 One-line diagram of the multi-machine system of Example 11.14.

4. Applying Euler's method to (11.49) and (11.50), calculate $\omega_i(t)$ and $\delta_i(t)$ for each generator.
5. Go to step 1 of the step-by-step simulation.

Example 11.14 *Time-domain simulation of a multi-generator system* Figure 11.28 shows the one-line diagram of a 60 Hz, three-generator, five-bus system. If we include the internal nodes of the generator, this network has a total of eight nodes. Let us simulate the behavior of this system when a fault occurs on the line connecting buses 6 and 7, close to bus 7. The tables below provide the system data in per unit on a 100 MVA basis.
Network data:

Line	R (pu)	X (pu)	$\frac{B}{2}$ (pu)
4–5	0.0	0.125	0.005
4–7	0.0	0.125	0.005
5–7	0.0	0.125	0.005
5–8	0.0	0.125	0.005
6–8	0.0	0.4878	0.005
6–7	0.0	0.333	0.005

Generator data:

Generator	X' (pu)	H (s)	V (pu)	P (pu)
G_1	0.08	10.0	$1.05 \angle 0°$	Slack
G_2	0.18	2.0	1.00	1.0
G_3	0.12	6.0	1.05	1.6

Load data:

Load	P (pu)	Q (pu)
L_7	3.0	1.0
L_8	2.5	0.3

A power flow computation on this data gives the following results:

$$\begin{pmatrix} \overline{V_4} \\ \overline{V_5} \\ \overline{V_6} \\ \overline{V_7} \\ \overline{V_8} \end{pmatrix} = \begin{pmatrix} 1.05\angle 0° \\ 1.0\angle -8.25° \\ 1.05\angle +2.83° \\ 0.958\angle -12.16° \\ 0.941\angle -21.15° \end{pmatrix} \text{pu} \quad \begin{pmatrix} P_{e,1} \\ P_{e,2} \\ P_{e,3} \end{pmatrix} = \begin{pmatrix} 2.9 \\ 1.0 \\ 1.6 \end{pmatrix} \text{pu} \quad \begin{pmatrix} Q_1 \\ Q_2 \\ Q_3 \end{pmatrix} = \begin{pmatrix} 1.450 \\ 0.695 \\ 0.790 \end{pmatrix} \text{pu}$$

From the complex power produced by each generator and its terminal voltage, we can calculate using (11.51) and (11.52) the current it injects in the network and its internal emf:

Generator	$\bar{I}$ (pu)	$\bar{E}$ (pu)
G_1	$3.088 \angle -26.66°$	$1.181 \angle +10.78°$
G_2	$1.218 \angle -43.05°$	$1.139 \angle +0.84°$
G_3	$1.699 \angle -23.45°$	$1.155 \angle +11.94°$

Using (11.53) and the magnitude of the voltages, we can calculate the equivalent load admittances:

Load	g_l (pu)	b_l (pu)
L_7	3.269	1.089
L_8	2.824	0.3389

Combining these equivalent load admittances with the admittances of the lines and generators, we build the admittance matrices for the pre-fault, faulted, and post-fault conditions:

$$\mathbf{Y}^{(pre\text{-}fault)} = \begin{bmatrix} -j12.5 & 0 & 0 & +j12.5 & 0 & 0 & 0 & 0 \\ 0 & -j5.555 & 0 & 0 & +j5.555 & 0 & 0 & 0 \\ 0 & 0 & -j8.333 & 0 & 0 & +j8.333 & 0 & 0 \\ +j12.5 & 0 & 0 & -j28.49 & +j8 & 0 & +j8 & 0 \\ 0 & +j5.555 & 0 & +j8 & -j29.545 & 0 & +j8 & +j8 \\ 0 & 0 & +j8.333 & 0 & 0 & -j13.363 & +j3 & +j2.04 \\ 0 & 0 & 0 & +j8 & +j8 & +j3 & 3.269-j20.080 & 0 \\ 0 & 0 & 0 & 0 & +j8 & +j2.04 & 0 & 2.824-j10.369 \end{bmatrix}$$

$$\mathbf{Y}^{fault} = \begin{bmatrix} -j12.5 & 0 & 0 & +j12.5 & 0 & 0 & 0 \\ 0 & -j5.555 & 0 & 0 & +j5.555 & 0 & 0 \\ 0 & 0 & -j8.333 & 0 & 0 & +j8.333 & 0 \\ +j12.5 & 0 & 0 & -j28.49 & +j8 & 0 & 0 \\ 0 & +j5.555 & 0 & +j8 & -j29.545 & 0 & +j8 \\ 0 & 0 & +j8.333 & 0 & 0 & -j13.363 & +j2.04 \\ 0 & 0 & 0 & 0 & +j8 & +j2.04 & 2.824-j10.369 \end{bmatrix}$$

$$\mathbf{Y}^{post\text{-}fault} = \begin{bmatrix} -j12.5 & 0 & 0 & +j12.5 & 0 & 0 & 0 & 0 \\ 0 & -j5.555 & 0 & 0 & +j5.555 & 0 & 0 & 0 \\ 0 & 0 & -j8.333 & 0 & 0 & +j8.333 & 0 & 0 \\ +j12.5 & 0 & 0 & -j28.49 & +j8 & 0 & +j8 & 0 \\ 0 & +j5.555 & 0 & +j8 & -j29.545 & 0 & +j8 & +j8 \\ 0 & 0 & +j8.333 & 0 & 0 & -j10.363 & 0 & +j2.04 \\ 0 & 0 & 0 & +j8 & +j8 & 0 & 3.269-j17.080 & 0 \\ 0 & 0 & 0 & 0 & +j8 & +j2.04 & 0 & 2.824-j10.369 \end{bmatrix}$$

Since there are three generators, the reduced admittance matrices are 3×3. Using (11.58), we build these matrices for the pre-fault, faulted, and post-fault conditions:

$$\tilde{Y}^{pre-fault} = \begin{pmatrix} 0.821 - j4.397 & 0.538 + j1.951 & 0.453 + j1.062 \\ 0.538 + j1.951 & 0.387 - j3.590 & 0.322 + j0.701 \\ 0.453 + j1.062 & 0.322 + j0.701 & 0.269 - j2.546 \end{pmatrix}$$

$$\tilde{Y}^{fault} = \begin{pmatrix} 0.044 - j6.443 & 0.070 + j0.906 & 0.055 + j0.150 \\ 0.070 + j0.906 & 0.111 - j4.121 & 0.087 + j0.237 \\ 0.055 + j0.150 & 0.087 + j0.237 & 0.068 - j2.951 \end{pmatrix}$$

$$\tilde{Y}^{post-fault} = \begin{pmatrix} 1.040 - j4.108 & 0.644 + j2.075 & 0.270 + j0.415 \\ 0.644 + j2.075 & 0.438 - j3.538 & 0.222 + j0.415 \\ 0.270 + j0.415 & 0.222 + j0.415 & 0.145 - j1.305 \end{pmatrix}$$

Using a time step of $\Delta t = 0.001$ s, we simulate the response of this system to a fault that occurs at $t = 0.1$ s and is cleared at $t = 0.25$ s. Figure 11.29 shows how the differences between the rotor angles of generators 2 and 3 and the angle of generator 1 evolve. The rotor angle of generator 3 fluctuates more widely than the rotor angle of generator 2 because it is connected to the rest of the system through higher impedance lines. Figure 11.30 shows that, if the fault is cleared at $t = 0.26$ s, generator 3 goes unstable while generator 2 retains synchronism with generator 1.

11.5.10 Damping

When the rotor speed deviates from the synchronous speed, the magnetic flux linking the damper windings is no longer constant. This variation in flux induces currents in the damper windings, which create a torque that opposes the speed deviations and thus damps

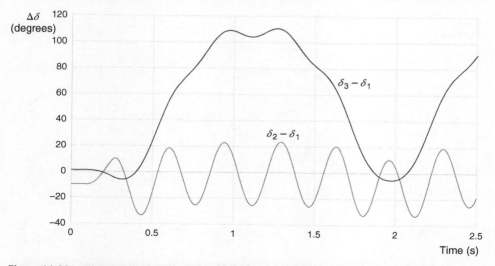

Figure 11.29 Time-domain simulation of a fault near the bus 7 end of line 6–7 of Example 11.14. The fault occurs at $t = 0.1$ s and is cleared at $t = 0.25$ s.

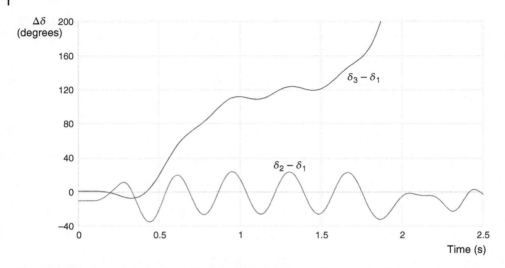

Figure 11.30 Time-domain simulation of a fault near the bus 7 end of line 6–7 of Example 11.14. The fault occurs at $t = 0.1$ s and is cleared at $t = 0.26$ s.

oscillations in the rotor angle. We can model this effect by adding a damping term to the swing Eqs. (11.49) and (11.50):

$$\frac{d\delta_i}{dt} = \omega_i(t) - \omega_{syn} \tag{11.62}$$

$$\frac{d\omega_i}{dt} = \frac{\omega_{syn}}{2H_i\omega_{i,pu}(t)}[P_{m,i} - P_{e,i}(t)] - D(\omega_i(t) - \omega_{syn}) \tag{11.63}$$

Example 11.15 *Multi-generator system with damping* Figure 11.31 illustrates how the relative angles of the generators of Example 11.14 evolve when the fault is cleared at $t = 0.25$ s as in the stable case but with a damping coefficient $D = 0.03/$s.

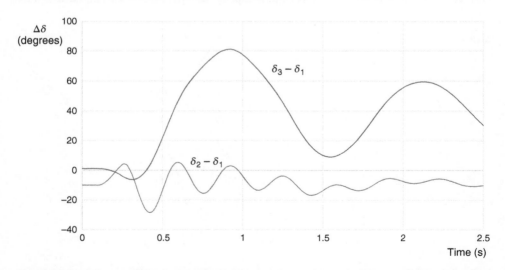

Figure 11.31 Time-domain simulation of a fault near the bus 7 end of line 6–7 of Example 11.14. The fault occurs at $t = 0.1$ s and is cleared at $t = 0.25$ s. The damping coefficient $D = 0.03/$s

11.6 Detailed Dynamic Models

In order to focus on fundamental concepts, our discussion of stability issues has so far been based on quasi-steady-state models and differential equations representing the mechanical behavior of the generators. Accurate stability assessments require the use of more detailed models that also capture how the electrical state of generators changes in response to a fault, how the loads change as the voltages in the network vary, and the effects of various control loops. The specialized treatises listed under Further Reading describe these models in great detail.

11.7 Power System Oscillations

The detailed models alluded to in Section 11.6 can be combined into an overall model of dynamic behavior of the power system, which can be written in the following compact form:

$$\dot{x} = F(x) \tag{11.64}$$

where x is a vector of electrical, mechanical, and control state variables, and $F(x)$ is a vector of non-linear functions. If we consider small deviations around an operating state x_0, we can replace (11.64) by a linear approximation:

$$\dot{x} = A(x_0)x \tag{11.65}$$

where the state matrix A depends on the operating state x_0. The eigenvalues of this matrix define the natural response modes of the system to small disturbances. Under normal circumstances, the real parts of all these eigenvalues are negative, which means that any disturbance is damped. However, under some circumstances, the real part of one of these eigenvalues can become marginally negative or even positive. This shift causes the emergence of poorly damped or growing oscillations in the system. These oscillations involve a periodic exchange of kinetic energy between generators: some generators accelerate while others slow down and then the process reverses at a frequency typically below 1 Hz. Such temporary differences in speed cause variations in rotor angles, which result in significant oscillations in the power flows on the lines connecting these generators.

11.8 Preventing Instabilities

Instabilities cause protective devices to detect a condition that might damage the equipment that they are monitoring. These devices react by disconnecting components from the rest of the system. While these outages prevent damage to the equipment, they can also cause other protective devices to react and disconnect more lines or generators. Such a cascade of outages brings about a rapid de-energization of all or part of the system, i.e., a blackout.

As discussed in the previous sections, instabilities are typically triggered by a sudden undesirable event, such as the disconnection of a line due to a fault, the loss of a generator or a rapid increase in load. Because such an event can happen at any time, operators must ensure that the power system is always in a state such that no *credible* contingency would cause cascading outages. A contingency is deemed credible if it involves the failure or disconnection of a single component. Losing simultaneously, or nearly simultaneously, two components is usually considered not credible because the probability of such an event is too low unless a common cause could lead to the disconnection of both. Operators must therefore continually perform a *contingency analysis*, which involves checking that no credible contingency would trigger cascading outages. Specifically, this means that:

- No outage should alter the power flows in such a way that another branch is overloaded and likely to be disconnected before remedial action can be taken as discussed in Section 11.4
- No outage should push the system beyond its stability limit, as discussed in Section 11.3
- No fault should trigger an electromechanical transient leading to an instability, as discussed in Section 11.5
- No outage should put the system in a state such that oscillatory modes are insufficiently damped or undamped, as discussed in Section 11.7.

Ideally such checks should be performed based on a detailed snapshot of the actual state of the system. Checking whether any contingency would cause a steady-state overload or undervoltage requires solving as many power flows as there are credible contingencies. Even for large systems, this can be done sufficiently fast to provide relevant information to the operator in real time. Performing contingency analysis for dynamic phenomena currently requires too much computing time to be practical online. To avoid these types of stability issues, operators rely on limits determined using extensive offline computations. When operators determine that a contingency would trigger an instability if it were to occur, they must take preventive actions. This means altering the state of the system by redispatching generation or adjusting other control variables.

Given that operators routinely perform contingency analysis, one may wonder why blackouts still occasionally happen. It is tempting to think that they are triggered by contingencies that were not considered credible and thus did not require preventive action. However, an analysis of major incident suggests that this is not the case. The root cause of most blackouts is typically a routine outage that should not create further problems but whose consequences are amplified by a failure in the information infrastructure of the power system. For example, a protective device may malfunction and mistakenly disconnect a second component thereby aggravating the situation. In other cases, operators did not have access to the information they needed to respond to a deteriorating situation.

Further Reading

Power System Stability and Control, 2nd Edition, by Prabha S. Kundur, Om P. Malik. McGraw Hill, 2022. The authoritative reference on power system dynamic modeling.

Power System Dynamics: Stability and Control, 3rd Edition, by Jan Machowski, Zbigniew Lubosny, Janusz W. Bialek, James R. Bumby. Wiley, 2020. A physically insightful analysis of many power system dynamic phenomena.

Power System Oscillations, by Graham Rogers, Springer, 1999.

Final Report - System Disturbance on 4 November 2006, UCTE, https://www.entsoe.eu/fileadmin/user_upload/_library/publications/ce/otherreports/Final-Report-20070130.pdf (accessed 30 November 2023).

Voltage Stability of Electric Power Systems, by Thierry Van Cutsem, Costas Vournas, Springer, 1998.

Power System Control and Stability, 3rd Edition, by Vijay Vittal, James D. McCalley, Paul M. Anderson, A. A. Fouad. Wiley-IEEE Press, 2019. Revised edition of the classic textbook by Anderson and Fouad.

Problems

P11.1   Consider the small system shown in Figure P11.1. Each of the lines is modeled as an inductive reactance of 0.75 pu. If the generator at Bus *A* maintains a constant voltage magnitude of 1.05 pu at its terminals, calculate the maximum active power that this system can supply if the load has a unity power factor, a 0.95 pf lag, and a 0.95 pf lead. What is the voltage at these critical loads?

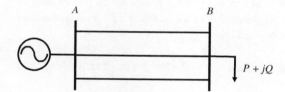

Figure P11.1   System of Problems P11.1–P11.6.

P11.2   Repeat Problem P11.1 for the case when only two lines are in service.

P11.3   The system of Problem P11.1 serves an active load of 1.5 pu. What is the lowest value of a lagging power factor that would make this load critical?

P11.4   What is the maximum unity power factor load that the system of Problem P11.1 can serve and maintain the magnitude of the voltage at bus *B* at 0.95 pu?

P11.5   Suppose that the system of Problem P11.1 is operating at the maximum unity power factor load calculated in Problem P11.4. What would happen if one of the lines is disconnected?

P11.6   The system of Problem P11.1 must serve an active power load of 1.20 pu at 0.95 pf lagging. What is the minimum amount of reactive power compensation needed to ensure that the system will not collapse if one of the lines is disconnected?

P11.7 The small system of Figure P11.2 is intended to supply a unity power factor load P_L. The generator at bus B supplies a fixed amount of active power P_B and maintains the magnitude of its terminal voltage at a constant value V_B. The generator at bus A supplies the balance of the active power load and maintains the magnitude of its terminal voltage at a constant value V_A. The parallel combination of the two lines connecting these two buses is modeled as an inductive reactance X. What is the theoretical maximum load P_L that this system can supply? What limitation on the generator at bus B would prevent this theoretical maximum to be reachable in practice?

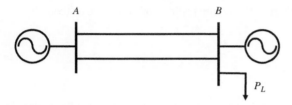

Figure P11.2 System of Problems 11.7–P11.10.

P11.8 Consider the small system shown in Figure P11.2. The two lines have the same impedance and each of them has a maximum rating of 75 MVA. The load is 100 MVA at unity power factor. If the generator at bus A has a significantly lower operating cost than the generator at bus B, determine how much power each of them should produce to avoid the possibility of a cascading outage. Assume that the flows of reactive power are negligible.

P11.9 Repeat Problem P11.8 for the case where one of the lines is rated at 100 MVA and the other at 50 MVA.

P11.10 Repeat Problem P11.8 assuming that there are three lines of equal impedance instead of two, each of them rated at 50 MVA.

P11.11 Repeat Problem P11.8 assuming that there are three lines of equal impedance instead of two, two of them rated at 50 MVA and one at 40 MVA.

P11.12 Repeat Problem P11.8 assuming that three lines connect buses A and B. These lines have the following characteristics:

Line	MVA rating	Impedance
L_1	40	Z
L_2	30	$2 \times Z$
L_3	50	$0.5 \times Z$

P11.13 A 50 Hz generator with an inertia constant $H = 3.5$ s is operating at rated power and a rotor angle $\delta_0 = 12°$ when a bolted fault occurs at its terminals. This fault is cleared after 150 ms Calculate the angle its rotor will have reached at that time. Assume that $\omega_{pu} = 1.0$ pu but that $\omega(t)$ varies with time.

P11.14 A synchronous generator has a synchronous reactance $X_S = 1.65$ pu and a transient reactance $X' = 0.25$ pu. Given that it operates at nominal voltage, calculate its steady state and transient emfs for the following conditions:
  (a) No load
  (b) Rated load at unity power factor
  (c) Rated load at 0.9 power factor lagging

P11.15 A synchronous generator with a transient reactance $X' = 0.30$ pu is connected to the rest of the system (modeled as an infinite bus) through a line modeled as an inductive reactance $X = 0.20$ pu. The magnitude of the voltage at the infinite bus is equal to its nominal value.
  Calculate the rotor angles corresponding to the stable and unstable operating points for the following conditions:
  (a) The generator delivers 1.0 pu of apparent power at 1.0 pf to the infinite bus
  (b) The generator delivers 1.0 pu of apparent power at 0.9 pf lagging to the infinite bus
  (c) The generator delivers 0.5 pu of apparent power at 0.9 pf lagging to the infinite bus
  (d) The generator delivers no apparent power to the infinite bus
  (e) The generator delivers 0.25 pu of reactive power to the infinite bus

P11.16 The 60 Hz generator of Figure P11.3 has a transient reactance $X' = 0.4$ pu. It is connected to the rest of the system, modeled as an infinite bus, by three lines of reactance $X = 0.5$ pu The circuit breaker at the infinite bus end of one of these lines is open. The voltage magnitude at the infinite bus is at nominal value, while the magnitude of the transient internal emf of the generator $E' = 1.3$ pu. This generator injects 1.0 pu of active power into the rest of the system. Calculate the critical clearing angle for a fault close to the generator end of the open-ended line. How fast should this fault be cleared to avoid instability if this generator has an inertia constant $H = 2$ s?

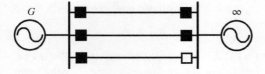

Figure P11.3 System of Problems P11.16–P11.18.

P11.17 Repeat Problem P11.16 for the case where the fault occurs at the generator end of one of the fully connected lines.

P11.18 Calculate the critical clearing angle for the system of Problem P11.16 assuming that the fault occurs at the midpoint of one of the fully connected lines.

P11.19 The 60 Hz generator shown in Figure P11.4. has a transient reactance $X' = 0.3$ pu and an inertia constant $H = 2$ s. It is connected to the rest of the system, which is modeled as an infinite bus, by three identical lines, each of which has an inductive reactance $X = 0.6$ pu. This generator delivers 1.2 pu active power at 0.85 power factor lagging at the infinite bus where the magnitude of the voltage $V_\infty = 1.05$ pu. Considering the worst location for a fault on one of the lines, will this system remain stable if the fault is cleared in 125 ms?

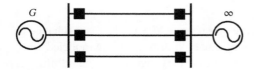

Figure P11.4 System of Problems P11.19–P11.21.

P11.20 Show that the system of Problem P11.19 would go unstable in the aftermath of a fault if one of the three lines were to be disconnected at both ends prior to the fault to carry out maintenance, while all other conditions remain the same.

P11.21 The 60 Hz generator shown in Figure P11.5 has a transient reactance $X' = 0.4$ pu and an inertia constant $H = 3$ s. The figure shows the impedances of the lines connecting it to the rest of the system, which is modeled as an infinite bus. This generator delivers 1.1 pu active power at 0.9 power factor lagging at the infinite bus where the voltage magnitude is $V_\infty = 1.05$ pu. Calculate the critical clearing angle and critical clearing time for a fault at location A.

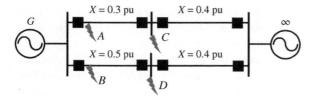

Figure P11.5 System of Problems P11.22–P11.25.

P11.22 Repeat Problem P11.21 for a fault at location B. Compare your results with those of Problem P11.21.

P11.23 The 60 Hz generator shown in Figure P11.5. has a transient reactance $X' = 0.4$ pu and an inertia constant $H = 3$ s. The figure shows the impedances of the lines connecting it to the rest of the system, which is modeled as an infinite bus. This generator delivers 1.1 pu active power at 0.9 power factor lagging at the infinite bus where the voltage magnitude is $V_\infty = 1.05$ pu. Calculate the critical clearing angle for a fault at location C.

P11.24 Repeat Problem P11.23 for a fault at location *D*. Compare your results with those of Problem P11.23.

P11.25 Write a computer program implementing Euler's integration method. Apply it to the integration of the following differential equation:

$$\frac{dx}{dt} = ax \quad ; \quad x(0) = 1$$

Use an integration time step $\Delta t = 0.001$. Demonstrate the correctness of your program by comparing your numerical results with the analytical solution of this equation $x(t) = e^{at}$ for $a = 2$ and $a = 3$. Calculate the relative error between the analytical and numerical solutions at $t = 1$. How small should the time step be to reduce this error to less than 0.1%?

P11.26 Write a computer program to carry out the time domain simulation of a one-machine infinite bus system. Use Euler's integration method with a timestep $\Delta t = 0.001$ s. Using this program and a trial-and-error approach, determine the critical clearing time and the critical clearing angle of the system of Problem P11.16. Compare your results with those you obtained using the equal area criterion.

P11.27 Use the program that you wrote for Problem P11.26 to check the results of Problem P11.17.

P11.28 Using the program that you wrote for Problem P11.26, check the value of δ_{crit} that you found in Problem P11.18 and determine the value of t_{crit}.

P11.29 Use the program that you wrote for Problem P11.26 to check the results of Problem P11.19.

P11.30 Use the program that you wrote for Problem P11.26 to check the results of Problem P11.20.

P11.31 Consider the system of Problem P11.19 and assume that one of the three lines has to be taken out of service for maintenance. How would you modify the *operation* of this system to ensure that it would remain stable in the event of a fault? Demonstrate your solution using the computer program that you wrote for Problem P11.26.

P11.32 Use the program that you wrote for Problem P11.26 to check the results of Problem P11.21.

Using time-domain simulation, we find $t_{crit} = 67$ ms and $\delta_{crit} = 38.51°$, which compares well with the critical clearing time obtained using the equal area criterion: $\delta_{crit} = 38.89°$ and the critical clearing time $t_{crit} = 68$ ms.

P11.33   Use the program that you wrote for Problem P11.26 to check the results of Problem P11.22

Using time-domain simulation, we find $t_{crit} = 115$ ms and $\delta_{crit} = 38.51°$, which compares well with the critical clearing time obtained using the equal area criterion: $\delta_{crit} = 56.16°$ and the critical clearing time $t_{crit} = 115$ ms.

P11.34   Use the program that you wrote for Problem P11.26 to check the results of Problem P11.23.

Using time-domain simulation, we find $t_{crit} = 114$ ms and $\delta_{crit} = 46.47°$, which compares well with the critical clearing time obtained using the equal area criterion: $\delta_{crit} = 47.04°$. Remember that we cannot calculate the critical clearing time using the equal area criterion when the generator can inject power during the fault because this injection makes the differential equations non-linear.

P11.35   Use the program that you wrote for Problem P11.26 to check the results of Problem P11.24.

Using time-domain simulation, we find $t_{crit} = 294$ ms and $\delta_{crit} = 92.80°$, which compares well with the critical clearing time obtained using the equal area criterion: $\delta_{crit} = 93.10°$. Remember that we cannot calculate the critical clearing time using the equal area criterion when the generator can inject power during the fault because this injection makes the differential equations non-linear.

P11.36   Write a computer program to simulate the effect of faults on the dynamics of a multimachine system. Test your program using the data of Examples 11.14 and 11.15.

12

Introduction to Competitive Electricity Markets

12.1 Overview: Why Competition?

Historically, the utility companies that ran most power systems were vertically integrated, which means that they were responsible for all development, operation, and commercial activities. This included building and operating generating plants, expanding and maintaining power networks, operating the transmission and distribution networks, and sending electricity bills to the consumers. These companies had a monopoly on all these activities over a service territory that had been allocated to them by a governmental authority. While vertically integrated monopoly utilities are still common, concerns about whether they are the best vehicle to achieve societal goals of economic efficiency, reliability, and sustainability have led some governments to restructure the industry and introduce competitive electricity markets. Because they do not have competitors, monopolies indeed do not have a strong incentive to be efficient. Introducing competition compels companies to make better decisions to thrive or survive. Over time, this improved efficiency should lead to lower costs for the consumers. To make competition possible, the various functions of a vertically integrated utility must be unbundled to separate those where competition is possible from those where a monopoly remains the logical choice. In particular, since building multiple transmission or distribution networks makes no sense from an economic or environmental perspective, network operation and network planning should remain monopoly activities. On the other hand, power plants can compete against each other to supply energy to consumers.

12.2 Fundamentals of Markets

Before delving into the organization and operation of electricity markets, it is useful to introduce some fundamental concepts from economics. Markets are locations where buyers and sellers meet to trade goods. While for millennia these were physical locations, in recent years they have increasingly become virtual. Besides providing a venue for executing transactions, markets also give buyers and sellers the opportunity to collect the information that they need to make decisions. When buyers go to the market (either in person or by browsing

Power Systems: Fundamental Concepts and the Transition to Sustainability, First Edition. Daniel S. Kirschen.
© 2024 John Wiley & Sons Ltd. Published 2024 by John Wiley & Sons Ltd.
Companion website: www.wiley.com/go/kirschen/powersystems

websites) they can see the different goods that are on offer, compare prices, and decide what to buy. Similarly, sellers can gauge the numbers of potential buyers as well as their interest in the various goods.

This ability to gather information makes it possible for the price of goods to settle at an economically efficient equilibrium. Let us examine how this equilibrium price arises using as an illustration the market for T-shirts. For simplicity, we will assume that all the T-shirts on the market are of the same type and quality. Suppose that a manufacturer of T-shirts has determined that its total cost for producing a quantity q of T-shirts is given by a cost function $C(q)$. The derivative of this cost function with respect to the quantity produced q is called the marginal cost function:

$$MC(q) = \frac{dC(q)}{dq} \tag{12.1}$$

Since the marginal cost is the derivative of the total cost, it depends only on the part of this total cost that varies with the quantity produced (e.g., the labor and the cotton) and not on the fixed part of the cost (e.g., the cost of building the factory). This function thus tells us how much it would cost to produce one more T-shirt. The marginal cost tends to increase with the amount produced because manufacturers incur extra costs (e.g., paying workers overtime) as their production volume increases.[1]

Figure 12.1 illustrates what the marginal cost function typically looks like for a particular manufacturer. Using this marginal cost function, this manufacturer can determine the production level that would maximize its profit. Suppose that it is able to sell any number of T-shirts for π dollars each. As Figure 12.1 shows, if it produces less than q^* T-shirts, each of them costs less to produce than the price at which it could be sold and thus generate a profit. On the other hand, each T-shirt beyond the first q^* costs more to produce than the price at which it can be sold. This manufacturer therefore has no incentive to sell more that q^* T-shirts if the price is π. In other words, the marginal cost function tells us how many T-shirts this manufacturer is willing to sell as a function of the price.

Let us now consider all the manufacturers selling the same T-shirt. Typically, each of them has a somewhat different marginal cost function because they are more or less efficient and purchase the raw materials at different prices. As Figure 12.2 illustrates for the case of two manufacturers, we can aggregate all of these functions into what is called the

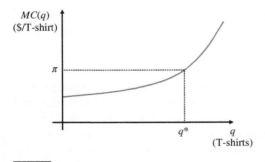

Figure 12.1 Typical marginal cost of production function and optimal production level.

---

1 The marginal cost of production should not be confused with the *average cost of production,* which is calculated by dividing the total cost of production $C(q)$ by the production volume q. Unlike the marginal cost, the average cost takes into account the fixed costs and initially decreases as the volume increases.

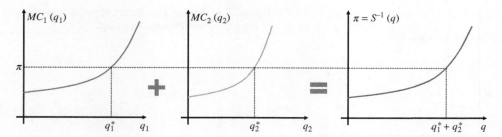

Figure 12.2 Aggregation of two marginal cost functions into an inverse supply function.

inverse supply function, which gives the price that would make these manufacturers willing to sell a quantity q of T-shirts:

$$\pi = S^{-1}(q) \tag{12.2}$$

Looking at this from the other direction, the supply function indicates the quantity that the manufacturers are willing to sell at a given price:

$$q = S(\pi) \tag{12.3}$$

Let us turn our attention to the buyers. People will buy a T-shirt only if the value or utility that they get from owning it is larger than the price they have to pay for acquiring it. While the amount of money that people are willing to pay for a T-shirt is a very individual decision, if we consider all the people who might buy T-shirts, it is clear that the number of T-shirts that they would be willing to buy decreases as the price increases. This is summarized by a demand function:

$$q = D(\pi) \tag{12.4}$$

Conversely, the inverse demand function indicates what the price should be for consumers to be willing to buy a certain quantity of T-shirts:

$$\pi = D^{-1}(q) \tag{12.5}$$

Figure 12.3 illustrates the typical shape of a demand or inverse demand function.

When they meet in a market, the suppliers' willingness to sell a commodity such as T-shirts is confronted with the consumers' willingness to buy these T-shirts. We can represent this mathematically by plotting the supply and demand functions on the same

Figure 12.3 Typical shape of an inverse demand function.

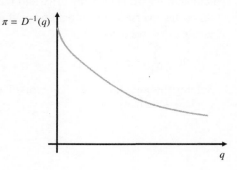

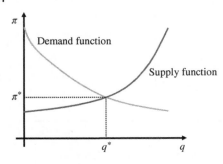

Figure 12.4 Market equilibrium at the intersection of the supply and demand functions.

graph, as shown in Figure 12.4. The intersection of these two curves defines a price π^* such that the quantity q^* that consumers are willing to buy is equal to the quantity that suppliers are willing to sell. Consumers will not buy more than this quantity because the utility that they would derive from these extra T-shirts would be smaller than the price they paid. They will also not buy less than this quantity because they could buy T-shirts for less than the dollar value they put on their utility. Similarly, suppliers will not sell more than q^* because they would have to produce these extra T-shirts at a loss. If they sold fewer T-shirts, they would forgo an opportunity to make a profit. All of these T-shirts will be bought and sold at the same price π^*, which is the price that "clears the market." Even consumers who most desire these T-shirts will not pay more because they can find suppliers who will sell it for that price. Conversely, even the most efficient manufacturers will not sell for less because they can find buyers willing to pay π^*. The market price is thus not just the amount of money coming out of our pockets when we buy a T-shirt. It is also the marginal cost of producing the last T-shirt sold on the market and the marginal utility that consumers place on that shirt. For these reasons, the market price is also called the marginal price.

It must be noted that this efficient market outcome occurs only under perfect competition, a condition that arises only when the market brings together enough sellers and buyers. If some of the sellers or buyers have too big a market share, i.e., sell or buy a very large proportion of the total quantity traded, they hold market power and may be able to artificially raise or lower the market price.

Example 12.1 *Market Equilibrium* If the inverse supply function for T-shirts is $\pi = 5 + \frac{q}{100}$ \$/T-shirt and the inverse demand function is estimated to be $\pi = 20 - \frac{q}{50}$ \$/T-shirt, calculate the market clearing price and the number of T-shirts that will be traded.

Since the inverse supply function and the inverse demand function intersect at the market clearing price, we have:

$$20 - \frac{q^*}{50} = 5 + \frac{q^*}{100}$$

Solving this equation, we get $q^* = 500$ T-shirts. Inserting this value in either inverse function gives $\pi^* = 10$ \$/T-shirt.

12.3   Wholesale Electricity Markets

Figure 12.5 illustrates the structure of a basic wholesale electricity market. Generating companies (Gencos) compete against each other to sell the electrical energy produced by the power plants it owns. The buyers in this market are load-serving entities (LSE), each of which supplies the consumers in a given region known as its service area where it also owns and operates the distribution network. Since the energy injected by the Gencos flows to the LSEs through the transmission network, one can say that wholesale competition takes place over the transmission network. An entity called the transmission system operator (TSO) manages this network to ensure that trades do not cause violations of operating limits and reliability issues. To maintain the fairness of the market, the TSO must be a separate entity, independent from the Gencos and LSEs.

While participants in these markets trade electrical energy (MWhs), the physical system is designed to deliver this energy as a continuous flow of electric power (MWs). To reconcile these commercial and practical perspectives, electricity markets define trading periods of typically one hour duration. Thus, when a Genco and an LSE trade a certain number of MWhs, they actually commit to, respectively, inject and extract this number of MW constantly during a given hour-long trading period.

Since flows of electricity are governed by the laws of physics, the energy sold by a given Genco cannot be directed through the network to the LSE that bought it. However, this does not cause any issue because MWhs are interchangeable, and the execution of the various trades can be monitored using meters at both ends.

Trading in wholesale electricity markets can be carried out either bilaterally or in a centralized manner. Bilateral trading is the traditional form of commercial transaction where a buyer and a seller interact directly to negotiate a price and a quantity. However, because the operational constraints of the transmission network must always be respected, the TSO may impose limits on the allowable trades. Centralized trading was developed to facilitate

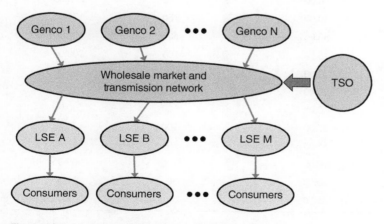

Figure 12.5   Structure of a wholesale electricity market.

the integration of the market with the operation of the transmission system. In a centralized electricity market, LSEs and Gencos do not trade directly with each other, but instead interact separately with the TSO, which takes on the role of market operator. Such market operate as follows for each trading period:

- Prior to the beginning of the trading period, Gencos submit to the TSO offers to sell energy. Each offer takes the form of a quantity/price pairs, for example "100 MW at 23 $/MWh." The capacity of each generating unit is typically divided into several such offers.
- The TSO ranks these offers in increasing order of price to create the supply curve.
- Similarly, LSEs submit to the TSO bids to buy energy. Each bid also takes the form of a quantity/price pair.
- The TSO ranks these bids in decreasing order or price to create a demand curve.
- The intersection of these supply and demand curves determines the market clearing price and the quantity transacted for this trading period.
- The offers whose price is less than the market clearing price and the bids whose price is higher than the market clearing price are accepted. The others are rejected.
- Gencos and LSEs whose bids and offers were accepted are then committed to inject and extract the corresponding amounts of power during that trading period.

Example 12.2 *Centralized Wholesale Market Clearing* Four Gencos and four LSEs participate in the electricity market of the little-known country of Syldavia. They have submitted the offers and bids shown in the table below to the TSO for the trading period between 10:00 am and 11:00 am on June 11.

Offers to sell			Bids to buy		
Genco	**Quantity (MW)**	**Price ($/MWh)**	**LSE**	**Quantity (MW)**	**Price ($/MWh)**
A	100	0.00	P	100	200.00
	100	30.00		100	175.00
	50	50.00		50	150.00
B	100	15.00	Q	100	200.00
	100	40.00		50	175.00
	50	90.00		50	10.00
C	100	0.00	R	100	200.00
	100	25.00		100	175.00
	50	65.00		50	15.00
D	50	10.00	S	100	200.00
	50	20.00		50	175.00
	100	35.00		50	5.00

The TSO ranks these offers to sell in increasing order of price to construct the supply curve shown in Figure 12.6. It also constructs the demand curve by ranking the bids to buy in decreasing order of price. The intersection of these two curves determines the market

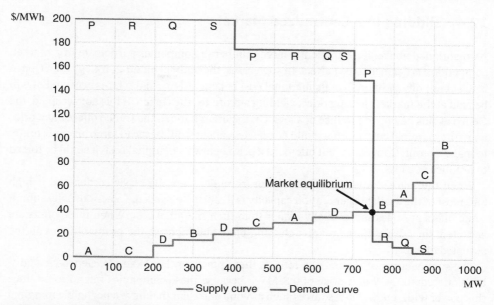

Figure 12.6 Supply and demand curves of Example 12.2. The letters indicate the companies that submitted the bids and offers.

equilibrium. For this trading period, the market clears at 700 MW and 40 $/MWh. The tables below show how each Genco and LSE fared in the market.

Genco	Accepted offers = quantity sold	Revenue
A	100 + 100 = 200 MW	200 × 40 = $8000
B	100 + 50 = 150 MW	150 × 40 = $6000
C	100 + 100 = 200 MW	200 × 40 = $8000
D	50 + 50 + 100 = 200 MW	200 × 40 = $8000
Total	750 MW	$30,000

LSE	Accepted bids = quantity purchased	Expenditure
P	100 + 100 + 50 = 250 MW	250 × 40 = $10 000
Q	100 + 50 = 150 MW	150 × 40 = $6000
R	100 + 100 = 200 MW	200 × 40 = $8000
S	100 + 50 = 150 MW	150 × 40 = $6000
Total	750 MW	$30,000

All the offers below the market price and the bids above the market price are accepted, while the other bids and offers are rejected. Only 50 MW of Genco B's offer of 100 MW at 40 $/MWh was accepted as this was the residual quantity needed to clear the market. This offer thus set the market price.

12.4 Bidding in a Centralized Market

As mentioned above, if a sufficiently large number of companies participate in a market, competition is deemed to be perfect. In such cases, the optimal bidding strategy for a Genco is to submit offers that reflect its marginal cost of generation. If an offer is accepted, it will be paid at the market clearing price, yielding a profit for the Genco. If it is not accepted, the Genco avoids having to produce at a loss. On the other hand, if the Genco offered at a price above its marginal cost, its offer could be rejected, and it would have to forgo an opportunity to make a profit. Conversely, if it offered at a price below its marginal cost, it could be forced to produce at a loss.

When the market is less competitive, for example when the expected demand is high and most of the available generation capacity is likely to be needed, Gencos usually submit some offers at prices substantially higher than their marginal cost. When these offers are accepted, they drive up the market clearing price and increase the profitability of all the accepted offers.

The supply curve of Figure 12.6 reflects these practices. For example, Gencos A and C both offered 100 MW at a price of 0 $/MWh because these companies intend to produce this power with renewable resources such as wind and solar that have a negligible marginal cost of production. The middle part of the supply curve rises gradually, reflecting differences in the heat rate and fuel cost of thermal power plants. The curve then rises more steeply, reflecting offers trying to take advantage of the reduced competitiveness.

While the price directly affects the Gencos' willingness to sell, it has a very limited influence on the amount that LSEs buy. Consumers indeed expect to be able to draw whatever power they need to get on with their lives and businesses independently of what happens in the market. A few of them may be willing to reduce their consumption if the price is too high. A few others can take advantage of a very low price to temporarily increase their demand. As the demand curve of Figure 12.6 illustrates, LSEs will therefore price most of their bids to guarantee that they will be above the market price. A few bids (e.g., LSE P's bid at 150 $/MWh) reflect the willingness of some consumers to curtail their demand. On the other hand, the bids of LSEs Q, R, and S between 5 and 15 $/MWh suggest that they would increase their load or store energy for future use if the price was sufficiently low. Some centralized electricity markets neglect the effect of the price on the demand and replace the demand curve by a vertical line at the value of the load forecast for the trading period. Economists describe such a vertical demand curve as being perfectly inelastic. In Example 12.2, replacing the demand curve of Figure 12.6 by a 750 MW forecast would not change the market clearing.

12.5 Variation of Market Price with Time

As we discussed in Chapter 2, the consumers' demand for electricity changes as a function of the time of day, time of the year, and weather conditions. Similarly, wind and solar generation also make the quantity offered dependent on these factors. The market equilibrium will therefore change with each trading period. In particular, the market price will increase when the demand rises and the supply decreases. Conversely, it will drop when the demand decreases or the supply increases.

Example 12.3 *Time-varying Prices* For simplicity, suppose that the Gencos of Example 12.2 submit the same offers for every hourly trading period of June 11. As discussed at the end of the Section 12.4, we will neglect the effect of the price on the demand curve and represent the demand by a vertical line at the value of the load forecast. The table below gives the values of these hourly forecasts.

Hour	1	2	3	4	5	6	7	8	9	10	11	12
Load	390	375	300	275	225	300	425	550	650	750	775	825
Hour	13	14	15	16	17	18	19	20	21	22	23	24
Load	810	775	750	750	825	875	925	825	750	650	450	425

Figure 12.7 illustrates the market equilibrium for hours 5 (minimum demand), 10, 19 (maximum demand), and 23. Figure 12.8 shows how the market price evolves over the course of that day.

Figure 12.9 illustrates the typical range of prices that arise in an actual electricity market, in this case ISO New England during the year 2022. This price duration curve shows the percentage of the number of hours in a year during which the price exceeds a certain value. Note the small number of hours during which prices reach extremely high values.

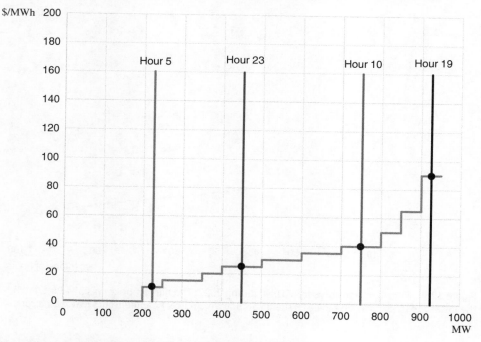

Figure 12.7 Market equilibrium at different hours of the day.

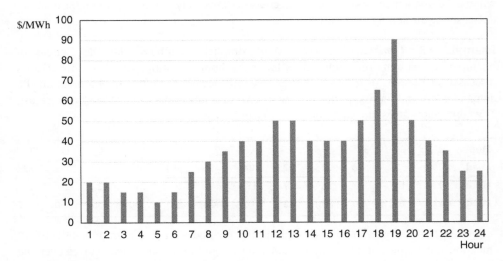

Figure 12.8 Evolution of the market price over the course of the day in Example 12.3.

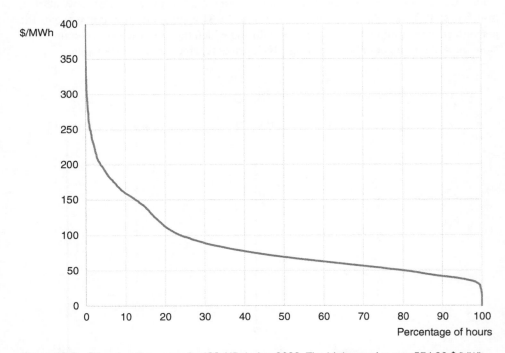

Figure 12.9 Price duration curve for ISO-NE during 2022. The highest price was 374.98 $/MWh, the lowest 14.02 $/MWh, and the average 85.56 $/MWh. Source: Adapted from www.iso-ne.com.

12.6 Effect of Transmission Capacity Limits

So far, we have assumed that at each trading period the market equilibrium is determined by the intersection of supply and demand curves that combine the offers and bids of all the Gencos and LSEs. Each generator then produces power according to its accepted offers and the market price is applied uniformly to all the energy transacted. However, the generation pattern resulting from these accepted offers would often cause power flows that violate operational limits on the transmission network. Since this is unacceptable, the market clearing process must be modified to satisfy these constraints. Essentially, some cheaper offers at the "wrong" location must be rejected and replaced by more expensive offers in the "right" place. A consequence of these adjustments is that the market price is then no longer uniform. Instead, the price that Gencos receive, and that LSEs pay, depends on where they inject or extract power in or out of the grid. This practice is called locational marginal pricing.

Example 12.4 *Locational Marginal Pricing* The electricity market of Syldavia that we introduced in in Example 12.2 is divided into two regions connected by a transmission line rated at 50 MW. Gencos A and D are located in the Western region, while Gencos B and C are located in the Eastern region. These four Gencos have submitted the same offers as in Example 12.2. The demand is modeled as an inelastic 750 MW load, 225 MW of which is in the Western region and 525 MW in the Eastern region. Figure 12.10a illustrates what would happen if the 50 MW limit on the transmission line was ignored when clearing the market. Generation would then exceed the load in the Western region, and 175 MW would flow from West to East on the transmission line. Since this exceeds the 50 MW rating of this line, generation in the Western region must be reduced by 125 MW and increased by the same amount in the Eastern region. Referring to Figure 12.6, we see that this will require rejecting D's 100 MW offer at 35 \$/MWh and curtailing A's offer at 30 \$/MWh to 75 MW. To compensate, in the Eastern region, the remaining 50 MW of B's offer at 40 \$/MWh must be accepted, as must C's 50 MW offer at 65 \$/MWh and 25 MW of B's offer at 90 \$/MWh. Figure 12.10b illustrates the generations and flow that result from this improved market clearing.

Another way of approaching this problem is to observe that the transmission constraint splits what we treated as a single market into two markets, one for the Western region and one for the Eastern region. In the Western market, A and D compete to supply the 275 MW demand (225 MW of local load plus the 50 MW outflow on the transmission line). In the Eastern market, B and C compete to supply the 475 MW demand (525 MW of local load minus the 50 MW inflow on the transmission line). As Figure 12.11 illustrates, the Western market clears at 30 \$/MWh, while the Eastern market clears at 90 \$/MWh. Generators in the Western region thus collect less revenue for each MW that they produce than their counterparts in the Eastern region. The transmission constraint also reduces the amount of energy that they can sell. Conversely, consumers in the Eastern region pay more for electrical energy than consumers in the Western region because the transmission constraint prevents them from accessing the cheaper Western generation.

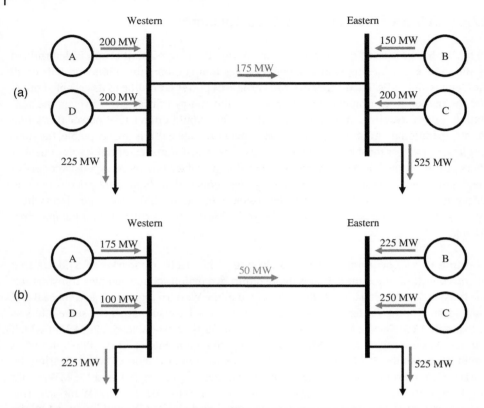

Figure 12.10 Generations and flow in the Syldavian electricity: (a) If the transmission constraint is ignored; (b) if the transmission constraint is incorporated in the market clearing process.

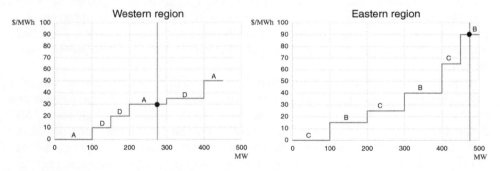

Figure 12.11 Market clearing in the Western and Eastern regions of the Syldavian electricity market.

Because the electricity market of the previous examples comprises only two zones connected by a single transmission line, it was relatively easy to identify what must be done to optimally adjust the results of the market to ensure that they respect the transmission limits. Clearing the market in actual systems where generators and loads are located at many nodes connected through multiple paths requires the solution of a linearized version of the optimal power flow problem described in Section 8.8. It can be shown that the prices of electrical energy at each bus are then given by the dual variables (Lagrange multipliers) associated with the corresponding power balance constraints.

12.7 Two-Settlement Markets

As we discussed in Section 6.5, substantial savings can be achieved if the operation of thermal generating units is scheduled over a day or more. Performing a unit commitment indeed makes it possible to optimize the number and type of generating units that are connected, to amortize their startup-costs, and to respect their operating constraints. To retain this benefit, many electricity markets operate both a day-ahead market and a balancing market.

On day D-1 of a day-ahead market, Gencos submit their offers to provide energy for every hourly or half-hourly trading period of day D. In addition to price-quantity pairs, these offers also specify the startup cost of the generating units and their operational constraints. At the same time, LSEs submit their demand curve for each trading period of day D. A complex calculation that combines a unit commitment and an optimal power flow then clears the market, i.e., determines the power to be produced or consumed by each entity at each location and every trading period, as well as all the locational marginal prices for all trading periods.

However, many things can happen between day D-1 and each trading period on day D: large generating units fail, wind and solar generation exceed expectations, load forecasts turn out wrong. To keep the system in balance, the power injections scheduled on D-1 must be incrementally adjusted up or down. Balancing markets operate shortly before real-time and are designed to determine these adjustments in an economically efficient, market-driven manner. In these markets, Gencos that can adjust their output quickly submit offers to increase or decrease their generation over 5- to 15-minute trading periods. LSEs that have the ability to alter their loads can also participate. The TSO determines which of these bids and offers it needs to call upon to keep the system in balance. The price of the most expensive bid or offer selected sets the real-time price.

Transactions cleared in the day-ahead market are settled at the day-ahead prices as if the energy actually produced or consumed was exactly equal to what had been scheduled. Any difference between the amounts scheduled on the day ahead and the actual values is settled at the real-time price.

Example 12.5 *Two-settlement Market* On June 10, the day-ahead electricity market of Syldavia cleared the hourly trading period 8 of June 11 at a price of 23 $/MWh at bus Patagonia. At this location during that trading hour Genco Blue was scheduled to inject 100 MW while LSE Orange was scheduled to extract 50 MW. The table below shows what Blue and Orange actually injected and extracted during each of the 15-minute trading intervals of the balancing market for hourly trading period 8. It also shows the real-time price at bus Patagonia, which reflects the bids and offers that the TSO accepted to maintain the power balance at that bus while respecting the operational constraints on the transmission network.

Trading interval	Injection by Blue (MW)	Extraction by Orange (MW)	Real-time price ($/MWh)
7:00–7:15	92	50	25.00
7:15–7:30	104	46	20.00
7:30–7:45	100	54	23.00
7:45–8:00	20	42	30.00

The table below summarizes the settlement of the day-ahead (DA) and real-time (RT) markets for Genco Blue for hour 8 of June 11. According to the standard accounting convention, parentheses indicate a negative value, i.e., an expense. The factors $1/4$ convert MW imbalances over a 15-minute period to a MWh value. Note the significant loss that Genco Blue incurred during the 7:45–8:00 trading interval because its injection fell significantly below the scheduled value.

Trading period	MWh	Price	Amount
DA: 7:00–8:00	100	23.00	$2300
RT: 7:00–7:15	$1/4 \times (92 - 100) = -2$	25.00	($50)
RT: 7:15–7:30	$1/4 \times (104 - 100) = 1$	20.00	$20
RT: 7:30–7:45	$1/4 \times (100 - 100) = 0$	23.00	$0
RT: 7:45–8:00	$1/4 \times (20 - 100) = -20$	30.00	($600)
		Total:	$1670

The settlement for LSE Orange is shown below.

Trading period	MWh	Price	Amount
DA: 7:00–8:00	50	23.00	($1150)
RT: 7:00–7:15	$1/4 \times (50 - 50) = 0$	25.00	$0
RT: 7:15–7:30	$1/4 \times (50 - 46) = 1$	20.00	$20
RT: 7:30–7:45	$1/4 \times (50 - 54) = -1$	23.00	($23)
RT: 7:45–8:00	$1/4 \times (50 - 42) = 2$	30.00	$60
		Total:	($1093)

12.8 Ancillary Services

While the TSO is responsible for reliably operating the transmission system, it does not own the resources needed to do so. It must therefore enter into contracts with Gencos and other entities that own these resources and can provide the necessary services. These services are called ancillary because they are accessory to the markets for energy but do not involve the recurring provision or consumption of a substantial amount of energy. The definition of these services varies from market to market, but typical examples include:

- *Contingency reserve*: As we discussed in Chapter 6, the system must be able to recover from a sudden significant imbalance between load and generation. This is possible only if enough generation capacity is held in reserve to respond to such contingencies. Since this reserve capacity cannot be used to produce and sell energy, generators that provide this service must be compensated for this loss of opportunity. Owners of battery energy storage systems, as well as large industrial consumers that are willing to disconnect part of their load at very short notice, can also provide this service.

- *Frequency control*: As we also discussed in Chapter 6, random fluctuations in load and generation cause small deviations in the system frequency. To maintain the frequency close to its nominal value, small adjustments must be made on a continuous basis to the power injections. On average, the net amount of energy required is close to zero. Generators operating on automatic generation control customarily provided this service. However, in recent years it has become a substantial source of revenue for battery energy storage systems.

- *Reactive power*: TSOs rely on the reactive power supplied by generators to control the voltages in the transmission network. However, as the loading capability diagrams of synchronous generators show (Figure 4.20), providing reactive power limits a generator's ability to supply active power and thus restricts its owner's opportunity to participate in the energy markets. Generators that inject reactive power must therefore be compensated for providing this service.

- *Black start capacity*: Before a large thermal generating units can be synchronized to the grid, a substantial amount of power is required to run its auxiliary systems (e.g., pumps and coal crushers) Under normal circumstances, this is just another load on the grid. However, in the event of a blackout, all generating units shut down and no power is immediately available to restart them. To restore the grid to a normal operating state, some generating units must therefore be able to restart without access to an external power source. These so-called black start generating units are typically hydro plants or small diesel generators whose power output is sufficient to restart the larger thermal units. These black start generating units are often not competitive in the energy markets. Since all power systems will at some point suffer a blackout, some Gencos must be compensated to maintain generating units with enough black start capacity.

12.9 Retail Markets

In some jurisdictions, the wholesale market has been complemented by a retail market. As Figure 12.12 illustrates, in these markets, companies called retailers purchase electrical energy in bulk on the wholesale market. These retailers then compete against each other to sell energy to individual consumers on the retail market. Since the retailers purchase electricity at the locations where distribution networks are connected to the transmission network and resell it to consumers who are located at the edges of the distribution network, one can say that retail competition takes place over the distribution networks. To ensure the fairness of the retail markets, these distribution networks are usually owned and operated by distribution companies (Discos) that are separate from the retailers.

12.10 Unbundled Industry Structure

The introduction of wholesale and retail electricity markets has required the breakup of vertically integrated, monopoly utilities. In the following paragraphs, we describe how the

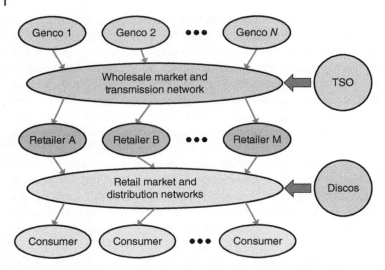

Figure 12.12 Combination of wholesale and retail electricity markets.

various functions that these utilities performed are typically unbundled between different types of companies. We also briefly discuss their sources of revenues.

A **generating company** (Genco) owns and operates one or more power plants. Some Gencos specialize in a specific generation technology (e.g., nuclear, wind), while others prefer to operate a portfolio of different types of plants. They collect revenues from the sale of energy and the provision of ancillary services.

Retailers buy electricity in bulk on the wholesale market and resell it to consumers at retail rates. To be profitable, these retail rates must be higher than the average price at which they purchased energy on the wholesale market. Retailers usually do not own significant physical assets.

A **transmission company** (Transco) owns and operates all the lines, substations, and other equipment that support the transmission of power from the generators to the distribution networks over a given geographical area. Transcos are thus regulated monopoly companies. They charge usage fees to the Gencos and retailers that inject and extract power in or out of their network.

A **distribution company** (Disco) owns and operates the distribution network serving a given geographical area. They are thus also regulated monopoly companies. Their revenues stem from fees that all consumers in their service area pay through their retailer.

The primary responsibility of the **transmission system operator** (TSO) is to maintain the reliability of the transmission system while operating the wholesale market in a fair and economically efficient manner. In some jurisdictions, it is called the independent system operator (ISO) or the regional transmission organization (RTO). These entities are usually non-profit and cover their costs by charging fees to users of the system. The only physical assets owned by TSOs are the control center and the communication equipment required for the day-to-day operation of the system. TSOs also often provide a forum where stakeholders can discuss and decide how to upgrade or expanding the transmission network.

Owners of storage systems can perform arbitrage in the wholesale electricity market, i.e., buy energy to charge their batteries when the price if low, and discharging these batteries to resell this energy when the price is high. They can also provide ancillary services to the TSO. Alternatively, they can operate in conjunction with renewable generators to firm up wind or solar energy and sell it at the best possible price.

Nowadays many aspects of power systems operate on the basis of free market economics. However, there is still a need for **regulators**. These government organization fulfill several functions. First, they ensure that the rules of the market are fair and encourage competition. Second, they set the rates that monopoly transmission and distribution companies can charge for the use of their networks. Third, they enforce the technical standards that all the types of companies described above must abide by to ensure the reliability of the power systems. In the United States, wholesale markets and transmission networks are regulated at the federal level while the various states regulate the retail markets and distribution networks.

Further Reading

Kirschen, D.S. and Strbac, G. (2018). *Fundamentals of Power System Economics*, 2e. Wiley.

Problems

P12.1 Suppose that the inverse demand function for jeans in a university town is $\pi = 100 - 0.01q$ \$/jean and the inverse supply function is estimated to be $\pi = 20 + 0.01q$ \$/jean. Calculate the market clearing price and the number of jeans traded.

P12.2 Given that the inverse supply function for T-shirts is $\pi = 15 + 0.01q$ \$/shirt and the demand function is $q = 700 - 10\pi$, calculate the market clearing price and the quantity traded.

P12.3 Economists estimate that the variable cost of production of electrical energy in the Bordurian electricity market is given by the following expression:

$$C(Q) = 20{,}000 + 500Q + 10Q^2 \text{ (\$) } \quad \text{for } Q \le 90 \text{ MWh}$$

$$C(Q) = 141{,}400 + 287.5(Q - 86)^2 \text{ (\$) } \quad \text{for } Q \ge 90 \text{ MWh}$$

where Q is in MWh.
They also estimate that the demand curve for electricity is given by the following expressions:
For the hour of maximum load: $Q = 100 - 0.00125\pi$ (MWh)
For the hour of minimum load: $Q = 55 - 0.001\pi$ (MWh)
where π is the price in \$/MWh.

  a. Sketch the supply and demand curves for this market.

b. Determine the following quantities at the market equilibrium for the hour of minimum load and the hour of maximum load:
- The quantity traded.
- The market price.
- The revenue collected by the producers.
- The total variable cost of production for all the producers.

P12.4 The supply curve for a commodity is given by the following expression:

$$\pi = 1000 + 20\,Q + 0.1\,Q^2 \ (\$)$$

where π is the price of the commodity and Q is the quantity.
Consider the following three demand curves for this commodity:
a. $\pi = 5000 - 20Q$
b. $\pi = 3000$
c. $Q = 100$

For each of these curves, determine the following quantities at the corresponding market equilibrium:
• The quantity traded.
• The market price.
• The revenue collected by the producers.
• The total variable cost of production for all the producers.
• The economic profit collected by these producers.

P12.5 Economists assess the consumers' sensitivity to the price of a commodity using a measure called the price elasticity of the demand, which is defined as the relative change in demand that results from a relative change in price:

$$\varepsilon = -\frac{\frac{dq}{q}}{\frac{d\pi}{\pi}} = -\frac{\pi}{q}\frac{dq}{d\pi}$$

Calculate this elasticity for each of the market equilibria of Problem P12.4.

P12.6 Given that the supply function for the widget market is $q = 0.2\pi - 40$ and the demand function: $\pi = -10q + 2000$, calculate:
a. The demand and the price at the market equilibrium.
b. The producers' revenue.
c. The producers' production cost.
d. The producers' profit.

P12.7 The TSO of the centralized electricity market of Borduria has received the following day-ahead bids and offers for the 10:00–11:00 trading period of January 25.

Offers to sell			Bids to buy		
Genco	Quantity (MW)	Price ($/MWh)	LSE	Quantity (MW)	Price ($/MWh)
Blue	300	10.00	Maple	250	300.00
Blue	100	30.00	Maple	200	250.00
Blue	50	60.00	Maple	50	20.00
Green	150	0.00	Oak	150	325.00
Green	200	20.00	Oak	100	275.00
Green	50	100.00	Oak	50	5.00
Orange	250	15.00	Fir	300	300.00
Orange	200	40.00	Fir	100	25.00
Orange	50	80.00	Fir	50	10.00
Purple	250	25.00	Larch	200	250.00
Purple	200	50.00	Larch	275	250.00
Purple	50	120.00	Larch	75	150.00

a. Build and plot the supply and demand curves for this trading period.
b. Determine the market price and quantity traded during this trading period.
c. Calculate the revenue collected by each Genco for this trading period.
d. Calculate the amount paid by each load serving for this trading period.
(Hint: Draw the supply curve neatly as you will need if for subsequent problems.)

P12.8 Instead of using a demand curve, let us assume that the electricity market of Borduria uses a price-insensitive forecast of the load to model the demand.

    a. Using the supply curve that you determined in Problem 12.7, determine the market price and the accepted offers if this load is forecasted to be 1200 MW for a given hourly trading period. Calculate the revenue collected by each Genco.

    b. If the load is divided among the LSE as shown in the table below, calculate how much each of them will have to pay for this trading period.

LSE	Fraction of system load
Maple	20%
Oak	30%
Fir	40%
Larch	10%

P12.9 Assuming that the supply curve you determined in P12.7 is valid for all the hourly trading periods of that day, draw a graph of the market price as a function of the time of day if the market clears at the quantities shown in the table below.

Hour	1	2	3	4	5	6	7	8	9	10	11	12
MW	950	800	725	625	800	950	1100	1325	1400	1500	1550	1600

Hour	13	14	15	16	17	18	19	20	21	22	23	24
MW	1625	1550	1425	1400	1500	1600	1725	1820	1775	1600	1300	1100

P12.10 Assume that the Gencos have submitted the same offers to the electricity market of Borduria that we introduced in Problem P12.7 and that a price-insensitive forecast of the load is used to model the demand.

    a. Calculate the market price, the accepted offers, and the revenues of each Genco for a load of 1300 MW.

    b. This market is divided into two regions connected by a single transmission line rated at 50 MW. Gencos Blue and Green are located in the Northern region while Gencos Orange and Purple are located in the Southern region. Calculate the market price and the accepted offers in each region, as well as the revenues of each Genco if the load is 500 MW in the Northern region and 800 MW in the Southern region.

    c. Discuss the differences between the results of parts a and b. Under these conditions, who would gain and who would lose if the capacity of the transmission line was increased?

P12.11 Repeat Problem P12.10 assuming that the load is 875 MW in the Northern region and 425 MW in the Southern region.

P12.12 Genco PowerMax injects power in Northern Borduria while LSE BPower extract power in Southern Borduria. Both participate in the electricity market of Borduria. At the close of the day-ahead market for November 5, for the 13:00–14:00 hourly trading period PowerMax had sold 300 MW at the northern price of 30.00 $/MWh, while BPower had bought 200 MW at the southern price of 25.00 $/MWh.

The electricity market of Borduria also operates a real-time market on 15-minute intervals that is used to settle imbalances between commitments in the day-ahead market and actual power injections. The table below shows the performance of these two companies as well as the real-time prices between 13:00 and 14:00 on November 5. Calculate the net revenue or expense for these two companies during this hourly trading period.

Trading interval	PowerMax injection (MW)	BPower extraction (MW)	Northern real-time price ($/MWh)	Southern real-time price($/MWh)
13:00–13:15	296	204	32.00	25.00
13:15–13:30	304	192	30.00	24.00
13:30–13:45	100	200	40.00	30.00
13:45–14:00	0	208	50.00	28.00

Index

Power Systems: Fundamental Concepts and the Transition to Sustainability, First Edition. Daniel S. Kirschen.
© 2024 John Wiley & Sons Ltd. Published 2024 by John Wiley & Sons Ltd.
Companion website: www.wiley.com/go/kirschen/powersystems